D1217533

GRIDLITE
GRiD

conran
DESIGN
guides

HOME OFFICE

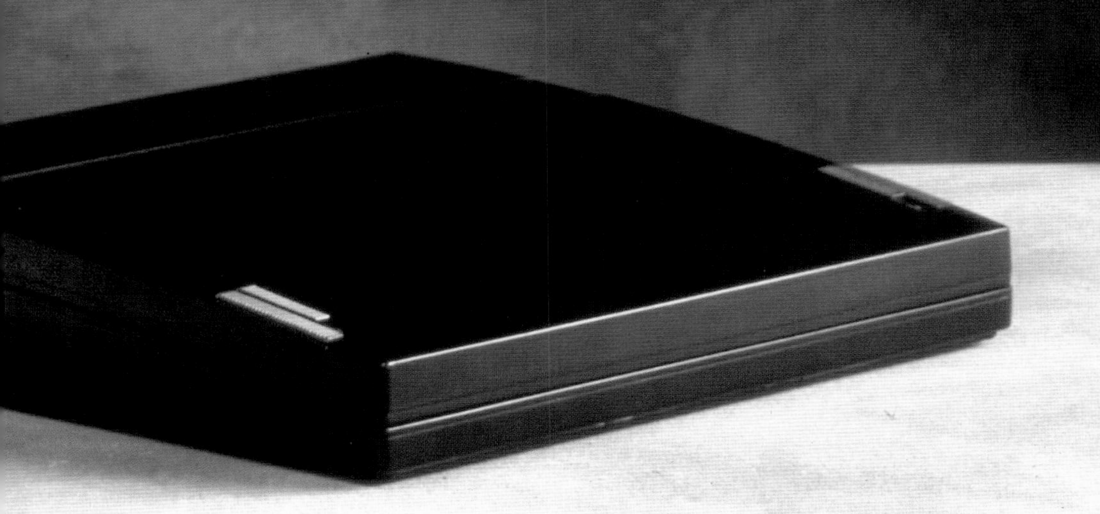

JEREMY MYERSON
& SYLVIA KATZ

VAN NOSTRAND REINHOLD
New York

Series Editor Joanna Bradshaw
Editor Mary Davies
Editorial Assistant Sally Poole
Design Paul Welti
Illustrator Cherrill Parris
Picture Research Kathy Lockley
Picture Research
Assistant Gareth Jones
Production Sonya Sibbons

16 15 14 13 12 11 10 9 8 7 6 5 4 3 2 1

Library of Congress Cataloging-in-Publication Data

Conran, Terence.
The Terence Conran design guides–home office / edited by Sir Terence Conran.
p. cm.
ISBN 0-442-30292-4
1. Office equipment and supplies. I. Title.
HF5548.C65 1990
651.2–dc20 89-70671
CIP

Published in the United States of America by
Van Nostrand Reinhold
115 Fifth Avenue
New York, New York 10003

First published in Great Britain in 1990 by
Conran Octopus Limited

Typeset by
Servis Filmsetting Limited
Printed by Wing King Tong

ACKNOWLEDGMENTS

The publisher thanks the following photographers and organizations for their kind permission to reproduce the photographs in this book:
2–3 Moggridge Associates; 5 top Wharmby Associates; 5 above centre Braun AG; 5 below centre Toshiba; 5 bottom Kartell; 6 Shona Wood/Conran Octopus (Charles Rutherfoord); 8 below Reproduction by permission of the Trustees of the Science Museum, London; 8 centre Jasper Seating Co; 9 courtesy of the Design Museum; 10 centre Courtesy of 'The Cabinet Maker' by Thomas Sheraton; 11 B.F.E.; 14–17 courtesy of the Design Museum; 18 Sotheby's, London; 19 above right Conran Design Group; 20 Rank Xerox; 22–23 Wharmby Associates; 24–25 Telefocus, a British Telecom photograph; 26 left Sotheby's, London; 26 right Mont Blanc; 27 above Henning Christoph; 27 below Bauhaus-Archiv; 28 above Henry Dreyfuss Associates; 29 above courtesy of the Design Museum; 30 above courtesy of the Design Museum; 30 below Pentel; 31 Telefocus, a British Telecom photograph; 32 above frogdesign/Photo Victor Goico; 32–3 below Plus Corporation, Tokyo; 33 above Wharmby Associates; 34–35 Braun AG; 36 left A.E.G.; 36 right courtesy of the Design Museum; 37 below Filofax; 38 above Henning Christoph; 38 below David Arky Photography, New York; 39 above Pictor International; 39 centre The Queens Museum, Phillis Bilick; 39 below Fritz Curzon; 40 Braun AG; 41 above Danese Milano; 41 below Danese Milano; 42 Conran Design Group; 43 Kartell; 44 below Plus Corporation, Tokyo; 46 above Kartell; 47 Fujii, Tokyo; 48–49 Toshiba; 50 Reproduced by permission of the Trustees of the Science Museum, London; 51 below Peter Williams; 51 above Henning Christoph; 52 above left Olivetti; 52 above right Olivetti; 52 below Reproduced by permission of the Trustees of the Science Museum, London; 53 above courtesy of the Design Museum; 54 above courtesy of the Design Museum; 54 below Olivetti; 55 Braun AG; 56 above Olivetti; 56 below Amstrad; 57 above courtesy of the Design Museum; 57 below Moggridge Associates; 58 above Plus Corporation, Tokyo; 58 below BIB Design; 59 above Toshiba; 59 below Canon; 60–61 Kartell; 62 above Sotheby's, London; 62 below Christie's, Geneva; 63 above SC Johnson & Son Inc; 63 below By Courtesy of the Board of Trustees of the Victoria & Albert Museum; 64 above the Museum of London; 65 above Sotheby's, London; 66 above Thousand and One Lamps Ltd; 66 below Ian McKinnell; 67 above Herman Miller Inc; 67 below Knoll International, The Knoll Studio Collection; 68 above left courtesy of the Design Museum; 68 above centre Clive Corless/Conran Octopus; 65 above right S.K.S. International; 68 below Kartell; 69 Kartell; 70 Gabrielle Vorreiter; 71 Ergonom; 72 Erco Lighting Ltd; 73–74 courtesy of the Design Museum; 75 above courtesy of the Design Museum; 75 below Knoll International, New York; 76 above Artemide/Photo Tom Vack; 76 below Danese Milano; 77 courtesy of the Design Museum; 77 courtesy of the Design Museum; 78 Braun AG; 79 above Popperfoto; 79 below courtesy of the Design Museum/Photo Peter Ogilvie;

The following photographs were taken specially for Conran Octopus by Simon Lee:

28 below, 29 below, 37 above, 44 above, 45, 46 below, 53 below, 65 below.

We would like to thank the following for their cooperation:
The Conran Shop
Cousins Design, New York
Design Museum
Environment
Bridget Kinally
Lisa Krohn and Tucker Veimeister, Smart Design, New York
Lefax
Plus Corporation, Tokyo
SCP
Seccose, Milan/Ideas for Living, London

Every effort has been made to trace the copyright holders and we apologize in advance for any unintentional omission and would be pleased to insert the appropriate acknowledgment in any subsequent edition of this publication.

AUTHORS' ACKNOWLEDGMENTS
The authors wish to thank all those manufacturers and designers who answered queries and searched through their archives, the supportive and professional staff at Conran Octopus and Sir Terence Conran for his personal interest and guidance.

NOTE TO READER
Names of objects and designers printed in roman or **bold** type denote that a photograph of the object or a biography of the designer can be found elsewhere in the book.

INTRODUCTION 6

1:TELEPHONES AND PENS 23

2:DESK ACCESSORIES 35

3:OFFICE MACHINES 49

4:FURNITURE AND LIGHTING 61

BIOGRAPHIES 72

INDEX 80

WORKING FROM HOME

Working from home is on the increase in Europe and North America. A convergence of new technologies, economic changes and social demands is dramatically reshaping the living patterns which have dominated much of the twentieth century. The nine-to-five routine which tied employees to a single location is being replaced by more flexible ways of working.

The home office became an attractive concept for both companies and employees during the 1980s. Corporations faced with steep rises in property prices and rents, and the runaway costs of heating and lighting, began to question the wisdom of running large corporate office environments. Employees, meanwhile, questioned the need to endure stressful daily commuting from the suburbs to work in crowded, often dirty, city centres. The home office offered a more cohesive lifestyle: work time could be more varied, involving evenings and weekends, to fit in better with the schedule of family life.

Demographic change, with women forming a higher proportion of the working population, and economic conditions, with more people becoming self-employed, have contributed to this shift. What has made the choice possible – and engineered major changes in the attitudes of corporate employers – has been the enormous advances in telecommunications, from answering and facsimile (fax) machines to personal computers, modems and electronic mail.

But choosing to devote a part of your home to a working office involves a series of important design decisions. No matter how popular the concept becomes, there are undeniable disadvantages to be overcome. These fall into two categories: physical and psychological.

How can you physically divide up your living space so that your home office has privacy? Do you want an entirely separate office or one which converts daily from the living area? Do you even have a choice?

A typical home office sited on a half landing: home-working is now a popular concept but major physical and psychological problems must be overcome.

And there are a host of other physical considerations. Where do you house your computer and printer? Should you invest in a separate phone line? Will you need an answerphone and a fax machine? What work surface should you use? How will you organize storage? What are the local authority laws on running a business from home?

It is the psychological disadvantages which are often harder to deal with, however. Home-workers, even those plugged into the most advanced electronic networks, complain of isolation, of not being able to share the work experience with others. The distinctions between work and leisure time are also torn down: lost is that ability to shut your mind to the job as you leap aboard the commuter train each evening.

But the design of the furniture and lighting, machines and accessories with which you equip your home office can go a long way towards solving the most fundamental physical and psychological problems.

It is argued that working from home makes you more professional, not less, simply because you are not in an environment custom-built for work and so have to redouble your efforts to overcome the obstacles. Artefacts and systems which enable you to organize your workspace in a more streamlined way and communicate with the outside world more efficiently are of paramount importance.

This book presents the milestone products which have paved the way during the twentieth century for the unprecedented sophistication of today's home office. Our selection is more than a buyer's guide for those contemplating a retreat from the communal office: it provides an insight into the work culture of modern times and the way it is now changing.

CHANGING WORK PATTERNS

Looking at the tall corporate office blocks that dominate the skylines of the world's major cities, it is easy to believe that offices began in commercial buildings. But, in fact, the earliest offices in the West were home offices. In early agrarian communities, farmers conducted transactions over the farmhouse kitchen-table, and this integration of home and 'office' space continued during the era of the wealthy burghers of medieval Europe.

The first banks were housed on the ground floors of private residences. Colonial traders and shipping magnates also converted part of their homes for the pursuit of business until expansion forced them to seek alternative accommodation.

Early swivel chairs for US office workers: the clerical industry mushroomed from 1880 onwards.

It was the Industrial Revolution in the early nineteenth century which stimulated the development of the modern commercial office. As the economy of first Britain and then other nations was transformed from an agrarian to an industrial base, as factories were built and cities developed, so the paperwork mounted up and the clerical industry was born. In the USA the number of clerks increased tenfold between 1880 and 1920.

A series of key inventions quickened the pace of office work in the late nineteenth century. In 1844 the first Morse telegraph had speeded up communications. This was followed by the introduction of the first practical typewriter in 1874, a Sholes and Glidden machine produced by E Remington & Sons, and the invention of the telephone by Alexander Graham Bell in 1876. Three years later Thomas Edison successfully developed the first tungsten-filament light bulb at Menlo Park, New Jersey.

Underwood No. 1 typewriter, 1897: this sturdy machine set the standard for the next 50 years.

Early Gestetner rotary duplicator dating from 1903: an exposed model which predates Raymond Loewy's restyling.

Architects joined designers and inventors in creating a new world of work. In 1884 William Le Baron Jenney designed what is arguably the world's first skyscraper, the Home Insurance Building in Chicago. By 1919, the year that the National Association of Office Managers was formed in the USA under the leadership of Frederick Taylor, the USA was leading the world in office practice. Taylor was an advocate of scientific management and carried out many time-and-motion studies to determine the most time-saving and cost-effective ways to organize office layout and procedures. Offices had become vast, complicated places peopled by large numbers of staff.

The look and layout of the commercial office has changed over the years, depending on the architectural fashion of the time. But whether cellular, open plan or bürolandschaft *(landscaped), the office has remained an enduring symbol of a century in which fewer and fewer people have actually made things and more and more people have processed information. Many developments in materials, furnishings and lighting can be traced directly to the needs of corporate and government employers and their staff.*

But even at its zenith many innovations developed for the commercial office – for example, ergonomically correct seating, uplighting and track lighting – quickly crossed the domestic threshold and so paved the way for the creation of effective home offices. And just as the late nineteenth-century inventions of Bell, Remington and Edison encouraged people to organize work away from the home, so a modern generation of designers and manufacturing companies are encouraging them to return home to work.

FURNITURE AND LIGHTING

Of all the items of equipment needed for a home office, the desk is the oldest and, in many aspects, the most unchanged. Despite a century of breathtaking scientific advances, a work surface is still a work surface – and many people still simply use the kitchen table as a desk.

Many contemporary pieces are reworkings of ideas from earlier centuries. For example, there is in the Victoria and Albert Museum in London an oak desk dating from c.1500 which has a hinged lid and a receptacle below for books. There are also references in fifteenth-century manuscripts to desks which revolve on a spiral column.

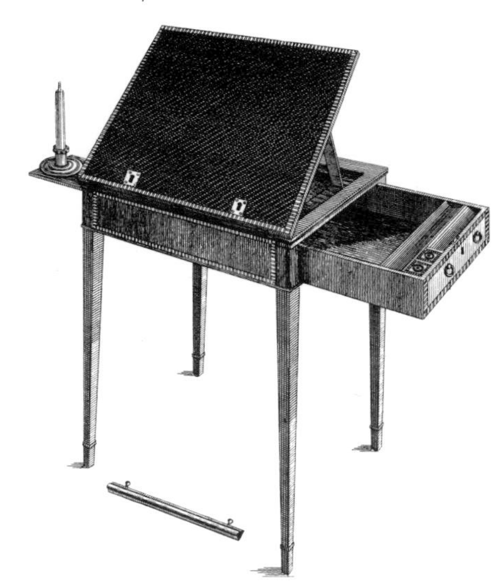

Technical ingenuity in English cabinet-making of the eighteenth century: a handsome reading and writing desk by Thomas Sheraton.

By the late eighteenth century English cabinet-makers such as Thomas Sheraton and Thomas Chippendale were showing a technical and creative ingenuity which in many ways has been unsurpassed. Their collection of cylinder writing tables, escritoires, secretaires, desks and bookcases, presents a wide variety of designs with decorative and functional elements perfectly combined. Reproductions of this work now abound. The originals are, of course, expensive antiques. Dating also from that time is the davenport, a small, sloping desk with a case of drawers below on castors. It was first made to order for a Captain Davenport by the firm of Gillow.

Davenport writing desk: once a drawing room essential, now an expensive antique.

In furniture Modernist design ideals in the early twentieth century centred on a simplification of form and the exploration of new materials. For example, chairs by Bauhaus participant Marcel Breuer in the 1920s and by Serge Chermayeff in the 1930s exploited the cantilever ability of tubular steel.

Later, after 1945, the special properties of newly developed plastics came into their own. Plastics brought a lightweight, space-saving potential to the home office. Furniture could be stacked or folded away, and new stacking storage and wall containers could be utilized. The Italian company Kartell, founded in 1949, was swiftest to see the possibilities. Its high-quality injection-moulded products – among them, Joe Colombo's storage trolley of 1970, Simon Fussel's Drawer system *of 1974, and Anna Castelli Ferrieri's stacking armchair of 1986 – have captured the imagination of design-conscious*

Boby trolley on castors designed by Joe Colombo, 1970: a clever exploitation of the storage potential of ABS plastics.

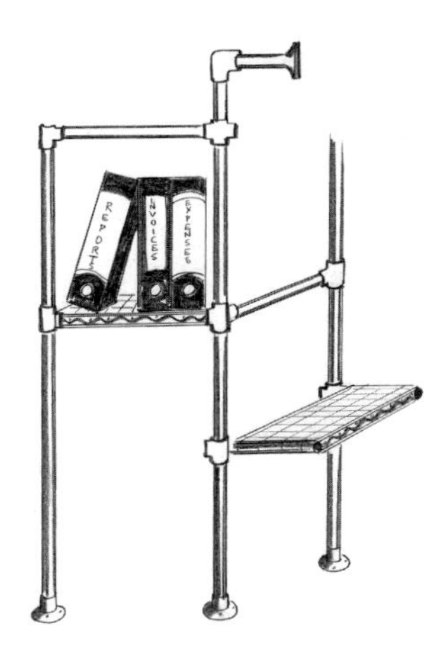

Ron Arad's 1980s Klee Klamp scaffolding system: economical construction for the home office.

home-workers, as did Florence Knoll's desks in the 1950s. Giancarlo Piretti's folding Plia chair (1969) *and Platone desk (1971), which combines ABS (Acrylonitrile Butadiene Styrene) plastics with chrome tubular-steel, are also classics of simple ingenuity, both for Castelli.*

Functionalist ideas have influenced many recent furniture pieces for the home office, most noticeably Israeli designer Ron Arad's high-tech system in the early 1980s, which used Klee Klamp scaffolding to create an office desk and shelving beneath a bed on stilts. But the latest thinking in home-office furniture is mining the rich seam of ideas from the time of Chippendale and Sheraton. Richard Sapper's 1989 Secretaire *for Unifor houses a personal computer but is inspired by eighteenth-century antiques and Massimo Scolari's meticulously crafted* Talo desk (1989) *also turns to Classicism for inspiration.*

However, if furniture for studies and libraries had a well-established repertoire by the late eighteenth century, developments in lighting for the home office belong very much to this century. In the years after Thomas Edison's creation of the first tungsten-filament bulb, this significant invention was refined by research and development to provide a brighter light.

By 1900 many offices were electrically lit, although they also relied heavily on natural light and had high ceilings to let in as much sunlight as possible. Meanwhile, in the home, Art Nouveau glassware designers such as Louis Comfort Tiffany in the USA and Emile Gallé and the Daum brothers in France were designing electric lights for wealthy, style-conscious customers.

The desk or task light has become one of the cult objects by which designers measure advances in theory and practice. Christian Dell and Wilhelm Wagenfeld at the German Bauhaus design school experimented with desk lights in the 1920s; the American pioneer consultants of the 1930s, such as Walter Dorwin Teague and Donald Deskey, did so too. But the most outstanding task light of the twentieth century was created in

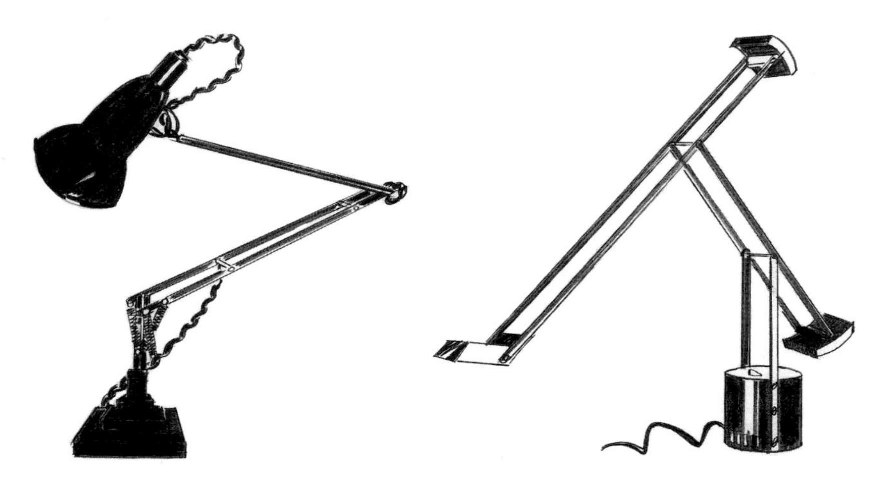

Evolution of the modern task light: (*left*) the Anglepoise, designed by George Carwardine for Herbert Terry & Sons, 1934; (*right*) the Tizio, by Richard Sapper for Artemide, 1972.

England in 1934, when a young automobile engineer, George Carwardine, collaborated with spring manufacturer Herbert Terry to design the Anglepoise lamp, which is based on the constant-tension principles of the human arm.

The Anglepoise has been scarcely altered to this day. Three years after it was launched, a Norwegian named Jac Jacobsen bought a patent, renamed the light Luxo *and introduced it first throughout Scandinavia and then, in 1951, into the USA with great success.*

The post-war years in home-office lighting have, however, been dominated by Italian designers who seized the opportunity presented by heat-formed plastics to create marvellous pieces of functional industrial sculpture. Milan in particular, the home of such progressive companies as Flos, Arteluce and Artemide, became the centre for new ideas. German designer Richard Sapper went to work there in 1958: he married Teutonic engineering precision to the sensuality of Italian design to create the Tizio light of 1972, a modern classic.

Since then, lighting technology has moved on apace with advances in low voltage and dimming fluorescent. These advances in light-source technology have been matched by even more ingenious designs of the fittings which house them, but little of this work is of direct relevance to the home environment. There is now a flourishing trade in reproductions of desk lamps from earlier this century, for example Mario Fortuny's circular-framed Table lamp *of 1900.*

TELEPHONES

Like the desk lamp, the telephone has fascinated designers. It is in many ways the most important instrument in the home office. As experienced home-workers will tell you, if you don't enjoy communicating over the telephone don't work from home.

Alexander Graham Bell, a speech therapist and self-taught physicist, was just 29 in 1876 when he uttered the first memorable words down a phone line, 'Come here Mr Watson. I want to see you.' The patent he took out that year on his invention has been perhaps the most profitable and profoundly important of all time.

Among the turn-of-the-century designs were Candlestick telephones *and the* Skeleton type *which was produced by the Swedish company L M Ericsson. These endured for many years until a radical change in telephone design was initiated first with the* Neophone *of 1929, which combined all the mechanisms within one unit, and then Ericsson's* Siemens model of 1931, *which set the pattern for much of this century.*

It was the geographical remoteness of Scandinavian communities depending heavily on telephone communication which prompted Ericsson's innovation. As automatic exchanges were installed in the 1920s, high development costs made it imperative to attract as many telephone subscribers as possible and the decision was made to redesign the instrument itself, using Bakelite to achieve a simpler, more streamlined form. The task fell to Jean Heiberg, a young artist with no engineering experience who had just returned from Paris to take up teaching at the National Academy of Fine Arts in Oslo.

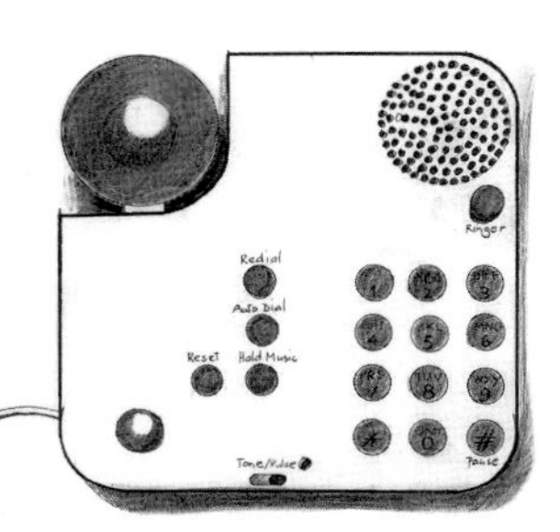

American Space-Tel telephone by Cousins Associates: one of the new designs of the late 1980s, following the liberalization of telecommunications.

Heiberg's 1931 telephone became a classic. Six years later the renowned American industrial designer Henry Dreyfuss performed a similar quantum leap for the Bell Company in the USA when he gave the Bell 300 telephone *a new aesthetic. Incredibly, these two designs remained unchallenged for two decades until Ericsson introduced the first one-piece telephone, incorporating handset and push-button dial, in 1956.*

In recent years, with the increased commercialization of telecommunications, the telephone has been the subject of much design innovation, some of it kitsch, some of it dedicated to improved ergonomics. It has ceased to be a standard utility and become a custom possession betraying the personality and social standing of its owner. Its integration and interaction with answering machines, fax machines, and personal computers via modems has opened up a new world for the home-worker. Telephones have become cordless and mobile, taking the office not only into the home but out onto the streets and into the car.

More technical breakthroughs are promised. British Telecom talks of holographic visiphones, while a prototype machine called Phonebook by American industrial designers Lisa Krohn and Tucker Veimeister suggests future developments. It combines address directory, answerphone and message-printout facilities with telephone – all within the familiar metaphor of the turning pages of a book.

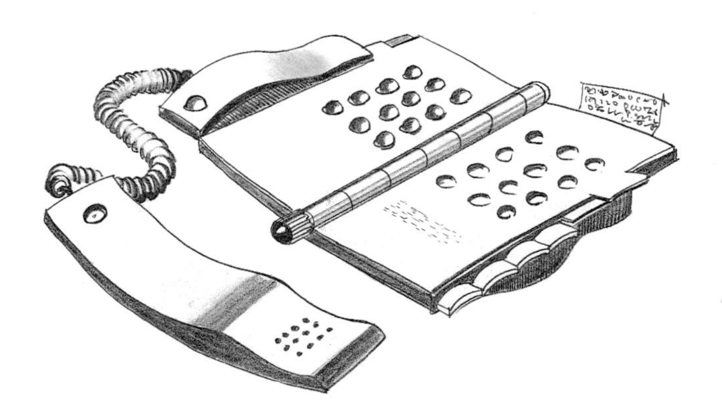

Phonebook by Lisa Krohn and Tucker Veimeister of Smart Design, New York, 1987: pointing the way to the future.

PENS AND PENCILS

Before the telephone was invented, the writing instrument was the most important tool in the home office, suggesting the class and refinement of the occupant. That still holds true to an extent today. Certain writing instruments have acquired the status of cult classics, such as the German Montblanc Meisterstuck 146, *designed in 1906, and the* Parker 51, *which was designed in 1939 to commemorate the American company's fifty-first anniversary.*

It was Parker which patented the first commercially manufactured self-filling fountain pen, using a piston to suck ink into a tube reservoir, in 1832. Earlier attempts to add ink reservoirs to quill pens had failed. The initial Parker fountain pen was a failure too. It leaked and was unreliable, and writers had to wait another five decades for the first successful model.

It was patented in 1884 by a New York insurance broker called Lewis Edson Waterman. The story goes that he decided to design a better fountain pen after he lost a deal to a rival broker when his pen refused to write. He was assisted by the availability of steel nibs and more suitable inks in the 1880s but mystified by the irregularity of the ink flow. Realizing that air should replace the ink which flowed out, he hit upon a capillary action with additional fissures in the feed to ensure an even flow. He called his first pen the Regular; it was entirely handmade with a vulcanite rubber barrel and wood decoration, and 500 were sold within a year.

Waterman left the insurance business and set up one of the world's most successful pen companies. A century later, in 1983, Waterman (by now a long-established French manufacturing company) celebrated its centenary with the launch of the luxurious Le Man 100 fountain pen.

Waterman has always traded on fashion and class. British prime minister Lloyd George signed the Versailles Agreement of 1919 with a solid gold Waterman. Indeed there has always

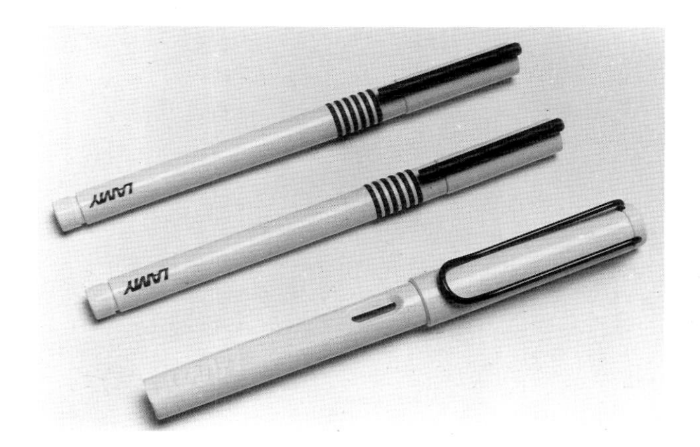

Writing accessories as style objects: German-made Lamy pens, designed by Wolfgang Fabian, 1982 became instant classics.

been sometimes special about the fountain pen. The competition may be less expensive but it is also less elegant.

The Biro ballpoint was patented by Laszlo Biro, a Hungarian artist and inventor, in 1938. Capillary action fed specially formulated quick-drying ink to a steel ball to ensure a consistent flow. The Biro's earliest use was by airmen drawing charts at high altitude during World War Two. The invention was later skilfully exploited by the Frenchman Marcel Bich, who used a tungsten-carbide ball and a clear acrylic plastics housing. Bich began marketing his Bic Crystal as the first disposable pen in 1958.

The world's first fibre-tip pen was developed in Japan by Pentel in 1963. It had a bamboo inner barrel which fed ink to the fibre tip. In 1966 the bamboo barrel was replaced by a solid tube of acrylic fibres but in other respects Pentel's much imitated original concept remains unchanged.

The pencil is the writing instrument least touched by time. Its origins date back to 1564 when a deposit of pure graphite was discovered in Cumbria in northern Britain. Blocks of the raw material were shaped into rods and bound with string to keep the fingers clean: as the pencil point was worn down, the string was unwound. Tubes and claws of metal were also *developed to hold the graphite. These crude instruments were the ancestors of today's propelling and clutch pencils.*

As supplies of pure graphite ran out in the seventeenth century, experiments began to make graphite composites and to glue the material to pine or cedar shapes. Famous European pencil manufacturers Kasper Faber and T & R Rowney began production in the late 1700s but it was not until the 1930s that a technique for securely bonding lead to wood finally reduced point breakage. Today, for all the microchip aids available, the pencil remains essential to the home office.

DESK ACCESSORIES

The cultural ritual of writing has, of course, stimulated the development of various accessories. Glass ink-bottles and wells date back to about 1750. By the early nineteenth century the exclusive Venetian glass-makers had a thriving trade in ornate desk sets accompanied by paper knives, pens, blotters, seals and so on. It was the Venetian glass-maker Pietro Bigaglia who exhibited the first decorative millefiori paperweights in Vienna in 1845.

By the early 1900s Art Nouveau glass-makers such as Louis Comfort Tiffany, René Lalique and the Daum brothers were

Stourbridge desk set in *millefiori* style, c. 1850: writing as a social and cultural ritual.

Crayonne Input range by Conran Associates, 1974: sturdy ABS containers streamline workspace.

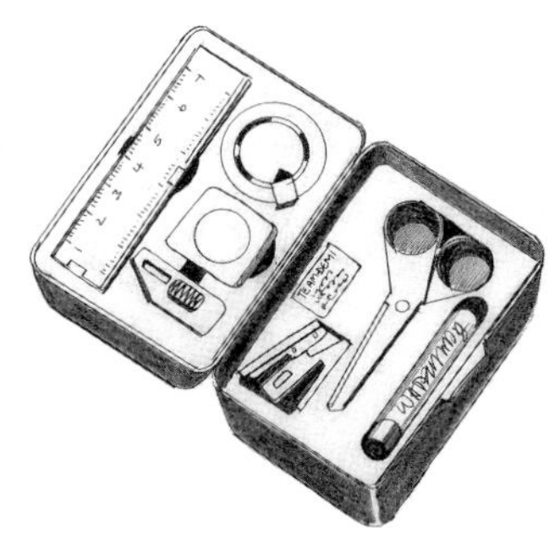

Team Demi mini office kit by Plus corporation, 1986: typical of the recent explosion of Japanese micro office design.

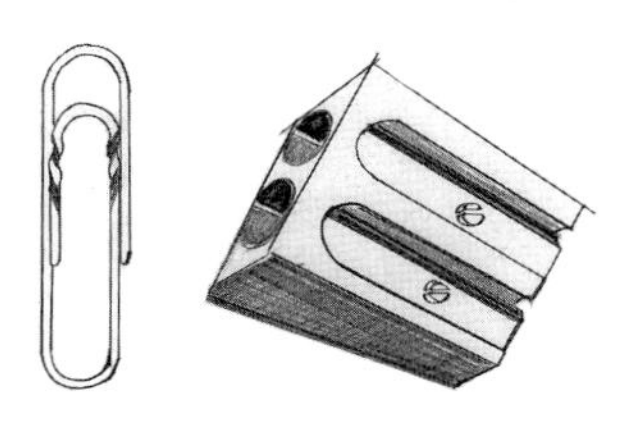

Enduring turn-of-the-century classics: the paperclip (Norwegian, 1899) and pencil sharpener (German, 1980).

producing desk accessories within this tradition, but a series of technological advances brought other innovations to the fore. At the giant AEG (Allgemeine Elektrizitäts-Gesellschaft) company in Germany, industrial-design pioneer Peter Behrens developed the first electric table fan *in 1908. And the influence of plastics has also become more and more significant: the ability to thermo-form or mould different shapes to create containers for all the paraphernalia of office life has been a great boon to the home-worker. The Crayonne Input range of 1974, designed by Conran Associates in Britain, indicated how such a versatile family of quality products could successfully streamline the workspace.*

By the 1980s, the Japanese in particular were exploiting the delicacy and subtlety of modern plastics by designing miniaturized interlocking office accessory sets. But some standard office accessories, it seems, cannot be bettered using new synthetics. The ubiquitous paperclip was designed in Norway in 1899 by Johann Vaaler. It has a brilliant simplicity which has survived endless attempts to supplant it. Much the same can be said for the reliability of the standard pencil sharpener, first developed by the TPX Bias Company in Germany in 1908.

First photocopier launched by Rank Xerox in 1959, more recent models have slimmed down in size.

ELECTRONIC MACHINES

However, it is the new technology which has enabled people to replicate the commercial office at home. Sophisticated electronics have proved to be just as liberating a force as mechanization was a century ago.

Two of the great industrial companies of the modern age, Olivetti and IBM, share much of the credit. Olivetti, founded in 1908 by Camillo Olivetti, has used such notable industrial designers as Marcello Nizzoli, Mario Bellini and Ettore Sottsass to give the typewriter portability and style. Olivetti's design leadership persuaded IBM president Thomas Watson to commission Bauhaus-influenced American designer Eliot Noyes to reshape the company's corporate identity for the 1950s. The IBM 72 Golfball electric typewriter, *designed by Noyes in 1961, remains a classic of its time.*

Other companies, though, have taken up the baton more recently. Apple Computers revolutionized home working with the introduction of the Apple Mackintosh *in the early 1980s, closely followed by the accessible and inexpensive Amstrad. Once huge and unwieldy, computers now sit on the desktop, an integrated element in the home office, and the typewriter is likely to become obsolete as new word processors and home computers, many with a sophisticated graphics capability, flood onto the market.*

Photocopiers too have been the subject of relentless development. Rank Xerox launched the first photocopier in 1959 but today's smaller, more efficient versions will sit on the desk (for example, Toshiba's 2810 tabletop photocopier*) and even operate in the hand (the* Copy-Jack *from Plus Corp in Japan). The first cordless photocopier was launched in 1989, the* Attache 11 *from Tetras in France, which works on batteries outdoors or plugged into the cigarette lighter socket in a car.*

But the greatest innovation, perhaps, has been the simultaneous development of the fax machine in Japan and the

The world's first cordless photocopier, the French-made Attache II by Tetras: you can use it in the street or car.

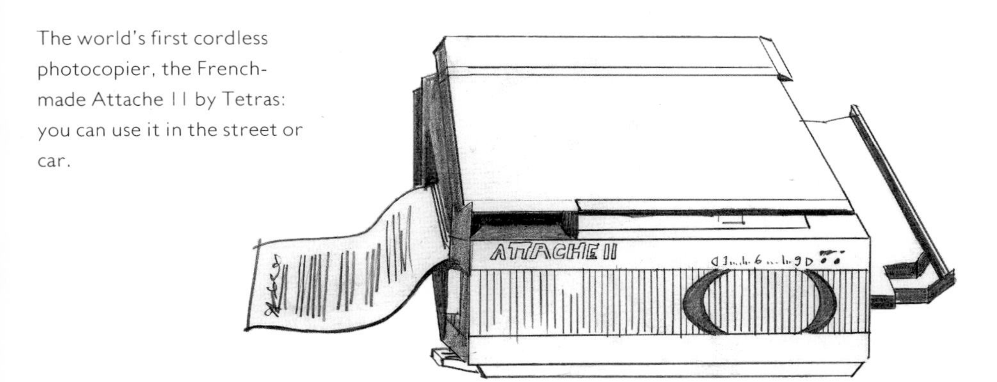

USA. When the first international patent was taken out in 1968 it took six minutes to transmit an A4 sheet. By 1976, the year of the second international patent, that time had been halved. Today the fax offers almost instant communication and is usually top of the shopping list for those who choose to work from home.

Why the fax should become so popular and other technological ideas, such as the video-telephone, should still await their time is one of the intriguing questions of home-office design development. Certainly we are in a period of transition. Some home-workers still have manual typewriters, others the latest graphics computers. Some still queue for stamps; others communicate by fax or electronic mail. Some consult paper-based Filofax *or* Rolodex *systems; others scroll up the information electronically on screen, or turn to their portable computerized diary.*

Some certainties remain. Computerization has not created the paperless home office: if anything, there is even more paperwork to store. But otherwise, as trends seem to be towards more home-workers and more technological breakthroughs, the home office is likely to be in transition for a long time yet. Designers clearly have a lot more important work to do.

1
2 ABC
3 DEF
4 GHI
5 JKL
6
7 PRS
8 TUV
*

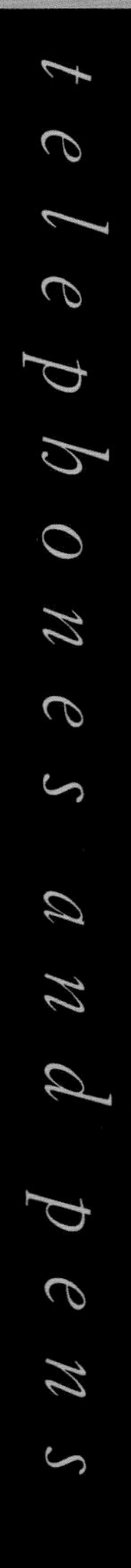

Written and verbal communication form the cornerstone of all home-office work, so that most fundamental design decisions concern which writing tools and telephones to choose. Technical innovation has been at the forefront throughout this century, bringing us in turn computerized and mobile phones, ballpoints and fibre-tip pens. But technological advance has been underscored by the way designers have created pens and phones which reflect the social status and cultural aspirations of those who use them.

Classics, of course, stand out – the turn-of-the-century Skeleton phone and the black Bakelite Neophone of 1929, the 1908 Montblanc Meisterstuck and 1939 Parker 51. But a recent wave of innovation, evident in new Japanese pen sets and the Taurus telephone, brings optimism for the future.

SKELETON TELEPHONE

made in Britain 1895

Just 11 years after Ericsson of Sweden combined transmitter and receiver to form the earliest telephone handset, the Skeleton telephone was introduced in Britain. An ingenious piece of industrial design embellished with Victorian decoration, it remained in production until the early 1930s. The Skeleton's attractively curved metal legs form the magnets of the generator, which is cranked up by the side handle. Copies abound and are popular among contemporary collectors.

CANDLESTICK TELEPHONE

British Post Office 1905

The Candlestick was developed just seven years before the first dial exchange opened in Britain. Its design – based on the need to keep the transmitter upright – has been so enduring that it is still on offer to phone users today. Importantly, the Candlestick made use of early modern materials: Vulcanite hard rubber and, later, Bakelite plastics. Speaking and listening were still regarded as distinct activities requiring their own apparatus.

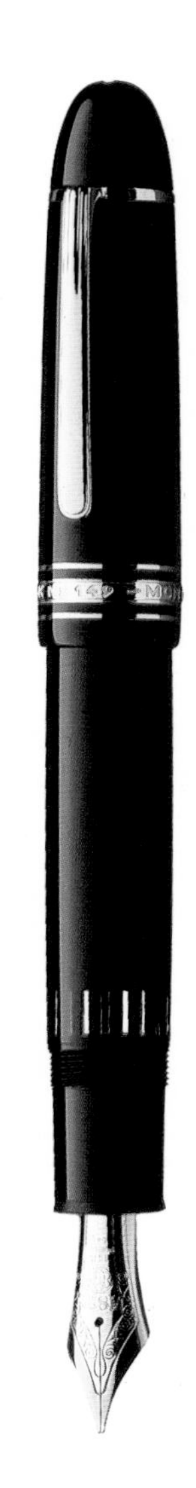

INKWELL
Louis Comfort Tiffany for Tiffany Studios
c. 1900
Tiffany was the USA's master of Art Nouveau textured surfaces. His popularity with the affluent is illustrated by this opalescent glass inkwell encased in a layer of bronze imprinted with stylized berries and leaves. The object shows how far writing accessories had become a measure of social status by the end of the nineteenth century.

MEISTERSTUCK 146 FOUNTAIN PEN
Montblanc c. 1908
A cult classic among writing aficionados and designers. The origins of the 146 date back to 1906, when a Hamburg banker, stationer and engineer teamed up to improve the fountain pen. The first Meisterstuck Montblanc was made by Simplo; today it is produced by Dunhill with an acrylic barrel. The 146 has cool, sleek looks and is pleasing to hold. It is now in the New York Museum of Modern Art and London's Design Museum.

NEOPHONE

made in Britain 1929

The Neophone was the first all-plastics handset model for use on the new automatic exchanges in Britain and was a major breakthrough in telephone design. Its now familiar form not only popularized the telephone but advanced the cause of plastics in product design. Most models were made of black Bakelite, but silvered and mock-walnut finishes were also produced and are rare and valued finds for collectors.

INK BOTTLE

Wilhelm Wagenfeld for Pelikan 1936–8

One of the most famous Bauhaus graduates produced the radical idea of making an ink bottle which was also an ink well. The 'W' shape represented Wagenfeld's signature and enabled the bottle to be stored upright in interlocking fashion, an imaginative, space-saving display solution for Pelikan, one of Germany's major suppliers of inks and pens.

BELL 300 TELEPHONE

Henry Dreyfuss for Bell
1937

When the American Bell Telephone Company commissioned ten designers to explore ideas for the future, Dreyfuss insisted on working 'from the inside out' alongside Bell engineers, rather than doing a purely stylistic exercise. His combined handset was made in metal and then, from the 1940s, in plastics.

DB 1001 TELEPHONE

Jean Heiberg for Ericsson 1931

If Henry Dreyfuss set the tone for American telephones for 30 years, Norwegian painter and sculptor Jean Heiberg did the same in Europe with his 1931 model for Swedish manufacturer Ericsson. In fact, Heiberg, working with engineer Johan Bjerkanes, was the true pioneer as his familiar design predates the Bell model by six years. Dreyfuss simply refined this design, which incorporates all mechanisms within one sculpted Bakelite body. Complete with pull-out drawer, Heiberg's telephone became the standard British GPO model, made in Britain by Siemens. It is still in demand today.

PARKER 51 FOUNTAIN PEN 1939

Kenneth Parker, Ivan D Tefft, Marlin Baker and Joseph Platt for Parker

Produced to celebrate Parker's fifty-first anniversary, the Parker 51 (*centre*) was machined from Lucite acrylic rod. Its slim, modern styling contrasts with Parker's Senior Duofold (*top*) of 1923 and the Centennial Duofold (*bottom*) of 1987.

BIRO BALLPOINT

Laszlo Biro 1938

The most important breakthrough in modern writing instruments was designed by Hungarian Laszlo Biro. He collaborated with his chemist brother George to feed a specially formulated fast-drying ink by capillary action to a steel ball which replaced the conventional nib. The Miles Martin Pen Company in England first manufactured Biro's wonder pen for World War Two airmen preparing maps at high altitude. Later the ballpoint became a consumer success, especially when Frenchman Marcel Bich took over the patent and produced the disposable Bic Crystal in 1958.

ERICOPHON

Ralph Lysell, Hugo Blomberg and Gosta Thames for Ericsson 1956

The world's first one-piece telephone. All functions are encased within a single housing made of SAN (styrene acrylonitrile). Dial, speaker and tone switch are all in the base; put it down on a surface and the circuit is cut off. Today it is a collector's item.

FIBRE-TIP PENS

Pental Research Unit for Pentel 1963

The fibre-tip pen is the only significant update on the ballpoint. It was developed in Japan, initially using a bamboo inner barrel to feed ink to the fibre tip. But in 1966 the bamboo barrel was replaced by a bundle of acrylic fibres which transferred ink to the tip by capillary action and reduced blotting. Since then Pentel has produced an enormous range of different colours and widths of tip, but the principle is unchanged: the fibre-tip pen has become an ubiquitous sight in offices all over the world.

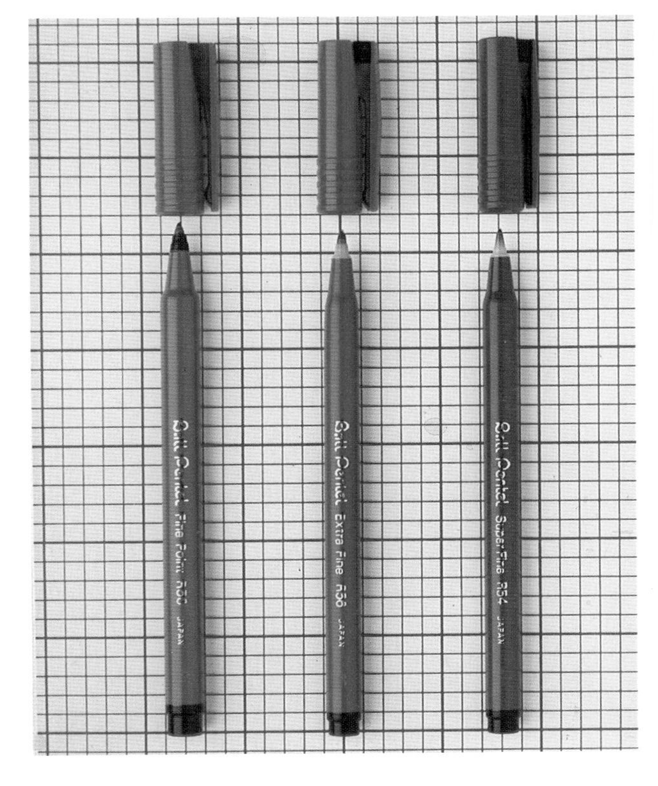

TRIMPHONE
Martyn Rowlands for STC 1964

Designed to be lightweight, compact and portable, the British Trimphone was intended as the second phone in the home, for use in home offices or on bedside tables. It was made of ABS plastics with tinted polycarbonate dial and cradle switch. The headset rests vertically north to south, rather than horizontally, on the phone, so breaking a long-held convention. Rowlands's design was criticized at the time for lacking the solidity of earlier models, but its simple, angular lines now symbolize the spirit of 1960s Modernism. It was the first phone of its kind to introduce an electric warbler, volume control and luminous dial. It won a British Design Council Award in 1966.

AEG CORDLESS PHONE

frogdesign for AEG 1985

Technological advance enabled the telephone to go cordless in the 1980s. This is one of the earliest models. Just as the phone was no longer tied to one place by the old physical constraints, so its styling avoided traditional domestic reference. This macho German product has a pseudo-military appearance drawn from field telecommunications equipment. It owes nothing to the more friendly, enduring home telephones of Dreyfuss or Heiberg.

Ps BAR PEN SET

Plus Corporation 1986

The Japanese set the latest trends in writing instruments in the mid 1980s by developing a generation of portable pen sets which are essentially fun, colourful, easy-to-use gift items. Plus Corporation's Ps Bar set is housed in an ABS (Acrylonitrile Butadiene Styrene) box with a lid which folds back to become a stand. The pens themselves are fibre tipped. The design combines educational and leisure aspirations for Japan's work-hard-play-hard young.

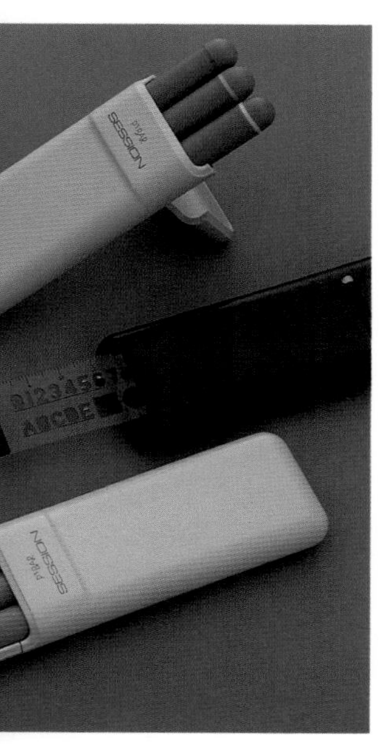

TAURUS TELEPHONE
Andy Davey for Browns Holdings 1988
A brilliantly sophisticated update of the famous black Bakelite telephone of the 1930s, the black Taurus is made of ABS and has a rubber cord. It is a stylish yet supremely functional home object with slide-out notepad drawer, number redial and memory facility, and a wall-hanging attachment. It won a coveted British Design and Art Director's gold award for its designer, Andy Davey of the London-based product-design consultancy Wharmby Associates, in 1989 and is widely regarded as a contemporary classic.

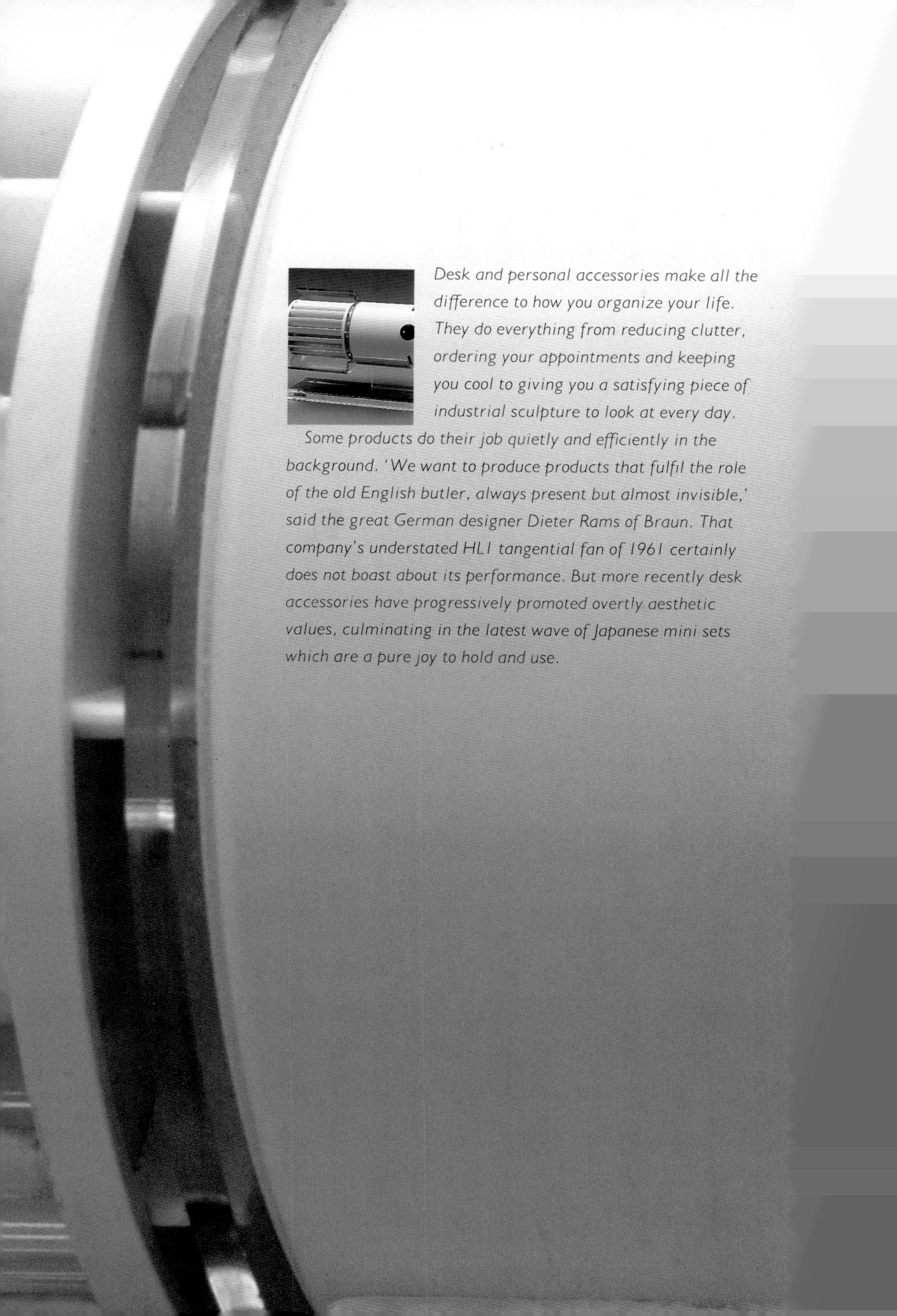

Desk and personal accessories make all the difference to how you organize your life. They do everything from reducing clutter, ordering your appointments and keeping you cool to giving you a satisfying piece of industrial sculpture to look at every day.

Some products do their job quietly and efficiently in the background. 'We want to produce products that fulfil the role of the old English butler, always present but almost invisible,' said the great German designer Dieter Rams of Braun. That company's understated HL1 tangential fan of 1961 certainly does not boast about its performance. But more recently desk accessories have progressively promoted overtly aesthetic values, culminating in the latest wave of Japanese mini sets which are a pure joy to hold and use.

ELECTRIC TABLE FAN

Peter Behrens for AEG

1908

The founding father of modern industrial design, Behrens developed a range of unadorned electrical products for AEG which emphasized technical function. His pioneering work as AEG's artistic adviser laid the groundwork for the Bauhaus in Germany and for the expression of Modernism in product design. The clean, coherent lines of this metal table fan with brass blades are an eloquent demonstration of his desire to tailor everyday goods to technological progress.

LEFAX PERSONAL ORGANIZER

J C Parker 1910

The Lefax looseleaf personal organiser was invented by a Philadelphia engineer by the name of J C Parker. He combined logic and craftsmanship to develop a portable ring-binder filing system for personal and professional use. The tactile Lefax range, now owned by Quarto, has always traded on high quality.

FILOFAX PERSONAL ORGANIZER

Norman & Hill 1921

One of today's most ubiquitous accessories in fact dates back to 1921 when Norman & Hill was formed. Originally a temporary secretary, company chairman Grace Scurr built up a strong clientèle in the clergy and military. But by the 1980s the versatile English Filofax had become an indispensable and portable home office.

ROLAM CALENDAR

Calendox early 1930s

The refinement of desktop accessories in the 1930s is demonstrated by this British-made circular calendar. There is fine detailing in the design of the phenolic resin housing and chrome frame. To amend the date you adjust the revolving celluloid window.

BANDOLERO DESK FAN

Diehl 1930s

This American electric desk fan was manufactured by Diehl, the electrical division of Singer, for the giant Sears mail-order house. It has a Bakelite housing and safety-conscious fabric blades, so doing away with the need for a guard. The idea of replacing potentially dangerous metal blades was subsequently developed further by substituting rubber blades.

CARVACRAFT DESK ACCESSORIES

Charles E Boyton for John Dickinson & Co 1948–51

These desk accessories are part of a range of 23 products moulded in mottled mock-amber cast phenolic. Dickinson, a British stationery company, had moulded shell casings and aircraft parts during World War Two and needed a peacetime use for its equipment. Designer Charles Boyton was noted primarily for his work in silverware and cutlery, but he adapted successfully to working in plastics. The use of cast phenolic enabled the moulding of thick sections, so giving the range its distinctive look. Carvacraft is sometimes mistaken for an example of 1930s styling.

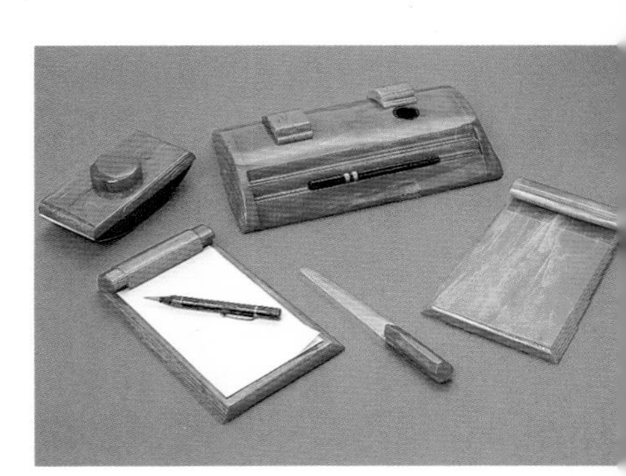

ARISTOCRAT STAPLER

Orlo Heller for E H Hotchkiss 1936

Claimed as 'the most beautiful stapler in the world', Orlo Heller's design reflected the influence of streamlining on American products. With its marblized phenolic resin housing, it is a piece of pure styling with no basis in function, the idea of swift movement being applied to a static desk accessory.

ROLODEX ROTARY FILE

Arnold Neustadter for Rolodex 1952

A classic American desk accessory which began life as a purely functional item, the Rolodex has become a cult object among designers. Made of heavy gauge steel, it rotates index cards which can be fixed in place at any point using a ball-bearing clutch mechanism called the 'Rolomatic'. Single and twin versions are still produced today.

HLI TANGENTIAL FAN

Reinhold Weiss for Braun 1961

The first desk fan with an enclosed tangential fan, this product has a clear family resemblance to other Braun appliances. All the mechanical parts are concealed in a simple, integrated housing. This appears to float on a transparent acrylic stand, which itself directs the airflow, while the circular shape of the machine body suggests the movement of the blades. Reinhold Weiss worked for Braun from 1959 to 1966 and is now based in Chicago. This classic design is now on display in the New York Museum of Modern Art as well as in the Design Museum, London.

EVERLASTING CALENDAR

Enzo Mari for Danese

1967

Italian willingness to experiment with new plastics in the 1960s is demonstrated by this ingenious calendar designed by Enzo Mari. The stand is made of ABS, the moveable indicators are printed PVC. The rational calendar folds at the relevant date into a simple, stylish right-angle on your desk – like a mini-road sign. The product was manufactured in Italian, English, French and German.

AMELAND PAPER KNIFE

Enzo Mari for Danese

1962

This Italian paper knife has a purity of form ergonomically geared to the task. You hold the knife, a twisted sliver of satin-finished stainless steel, at its wider end on one plane and the twisted form enables the easy opening of letters using the sharp end on another plane. Danese also produced a version in silver.

WALLPOCKET
Dorothee Maurer-Becker early 1970s
Made of thermo-formed ABS sheet, this wall-hung storage unit set out to solve the problem of clutter in the home office and workshop while ensuring that all the different-sized objects and accessories remained at your fingertips. The unit included 32 slots, pouches, hooks and clips to take the strain off the desktop. In retrospect the Wallpocket is more important as an idea than as a practical realization: the pouches became dust pockets which proved hard to clean. Maurer-Becker's solution is sculptural and imaginative, but the problem is still with us.

TABLE ASHTRAY

Anna Castelli Ferrieri for Kartell late 1970s

This patented ashtray was designed at a time when there were many more smokers about than now and set out to solve one of the anti-social aspects of smoking in the office. The main feature of the melamine design is that it instantly extinguishes cigarette stubs leaving no smoke or smell. As soon as a stub is inserted into one of the holes it goes out due to lack of oxygen. However, the contoured rim provides a convenient resting place for smouldering cigarettes. Anna Castelli Ferrieri, one of Italy's most outstanding talents in plastics, has been Kartell's art director since 1976. This ashtray won a coveted Compasso d'Oro prize for industrial design in 1979.

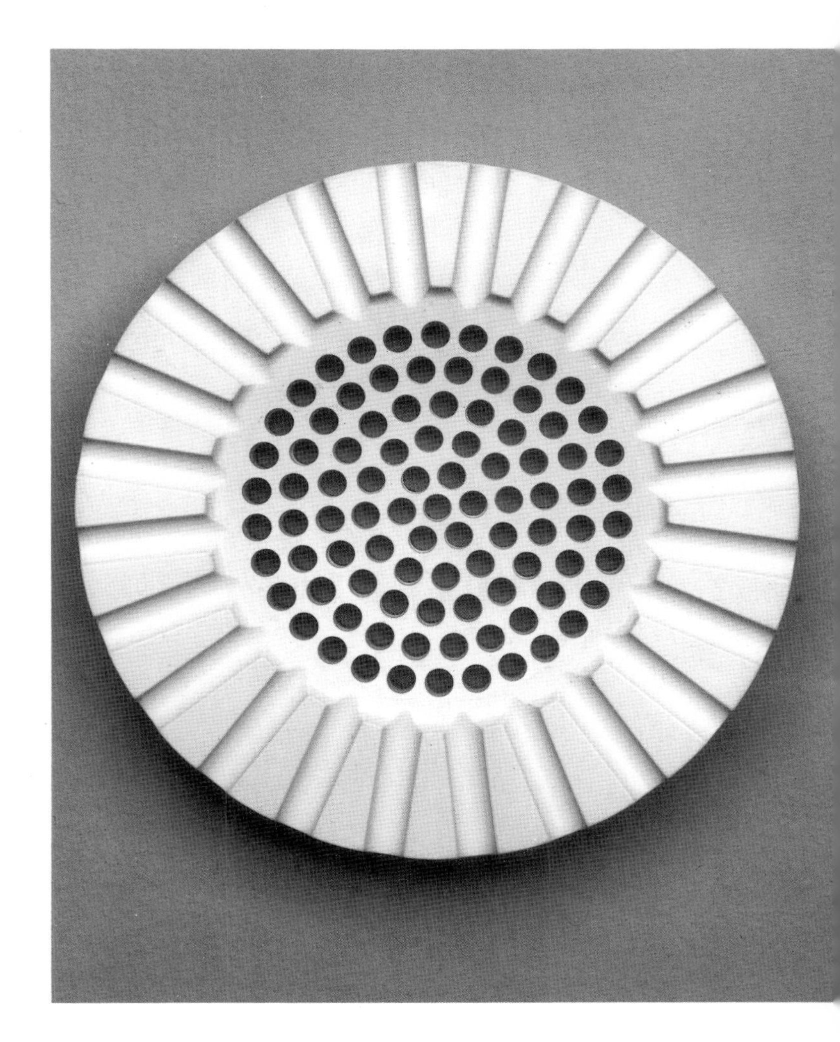

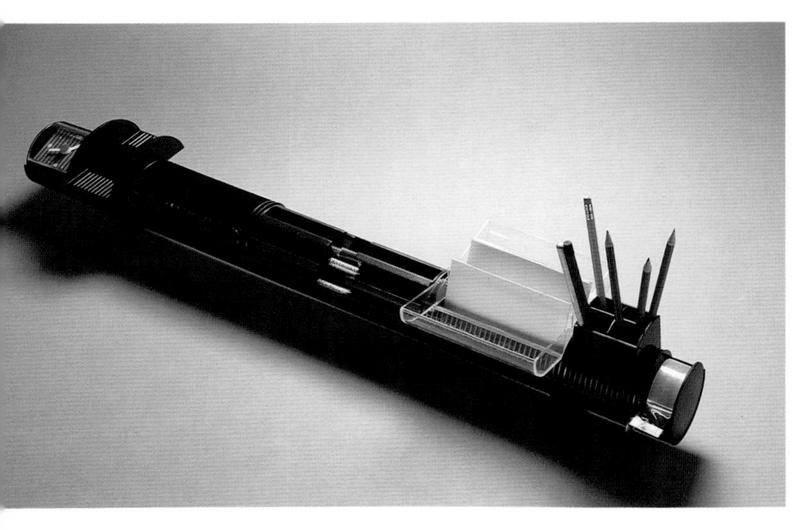

TABLE TRACK SYSTEM
Yaacov Kaufmann for Seccose 1989
This Italian design for an integrated system in plastics and aluminium by Yaacov Kaufmann, one of Israel's leading industrial designers, shows the huge advances since the early days of Bakelite desk accessories. The system includes clock, ashtray, revolving pencil-holder, paper tray, photo frame and tape dispenser.

FACTORY OFFICE KNIFE
Yoshihisa and Kohji Imaizumi for Plus Corporation 1986
A brilliant update on the Swiss Army knife by the Imaizumi brothers, the brains behind Japan's innovatory Plus Corporation, the much-imitated Factory was the first of a new generation of compact Japanese mini tool-kits in plastics casings. All its features fold neatly away, including scissors, stapler, staple remover, hole puncher, sticky-tape dispenser, magnifying lens, retracting tape measure and pincase.

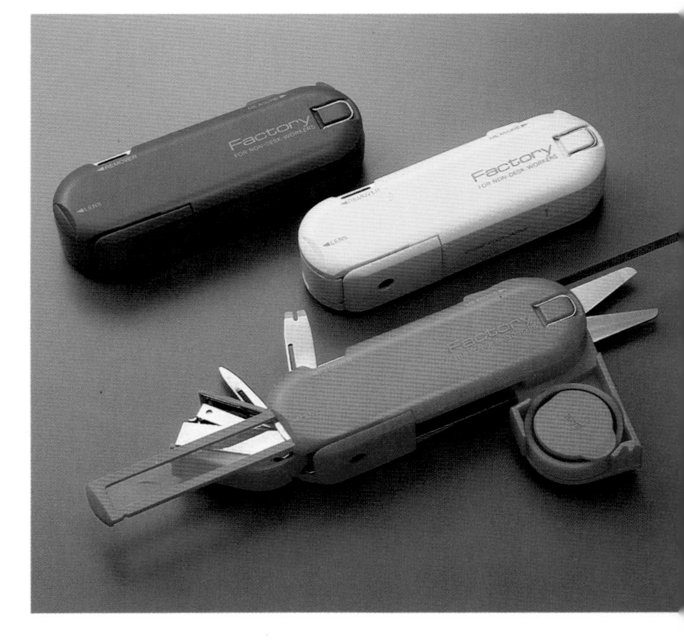

EL CASCO SHARPENER

Olave Solozabal 1920

The king of pencil sharpeners, this Spanish design became a cult object during the 1980s. Yet its origins stretch back to 1920, when Olave Solozabal, a manufacturer of firearms, first designed and produced the model in Eibar, northern Spain. It is made of steel and chrome, and has a lacqurered wooden handle and two tempered-steel cutting rollers which can be adjusted to four different grades. There is also a black rubber suction-base to secure the sharpener on the desk, a magnifying lens on top and a pull-out tray to remove shavings. The sharpener is hand finished, guaranteed for life and widely regarded as the world's finest.

SEGMENTI DESK SET

Michele de Lucchi and Tadao Takaichi for Kartell 1989

A strong, geometric solution in plastics and rubber, the desk set consists of what looks like segments of a giant extrusion. Turned at different angles the slanted planes become trays and containers.

PALETTACOM STATIONERY KIT

Bandai 1986

The Palettacom is quite unlike any other Japanese mini office tool-set. It draws its inspiration from a combination of ancient oriental puzzles, mind-bending origami and the colourful patterns of the De Stijl art movement. It is as though Rietveld had been asked to pack his entire drawing office into a tiny box. Made of acrylic and high-impact polystyrene, the set is an ingenious jigsaw puzzle in which the pieces turn out to be a hole puncher, tape measure, scissors, storage case, tweezers, paper knife, ruler, magnifying glass and stapler. The urge to play with the set is irresistible and may stop you actually doing any work!

FISSO DESKTOP ORGANIZER

Takenobu Igarashi for Fujii 1988

Office accessories enter the realms of abstract sculpture with this beautiful Japanese desk set. Fisso is an Italian name meaning 'firm, unalterable, fixed'. Takenobu Igarashi worked primarily in graphics before making the transition into product design. This coordinated plastics range, known as 'art stationery' in Japan, reflects today's greater emphasis on personal expression in the business world. It includes a postal scale, circular stamp pad, triangular utility case and scissors, pens, paper knife and letter trays. It has been manufactured in colours too.

BD-

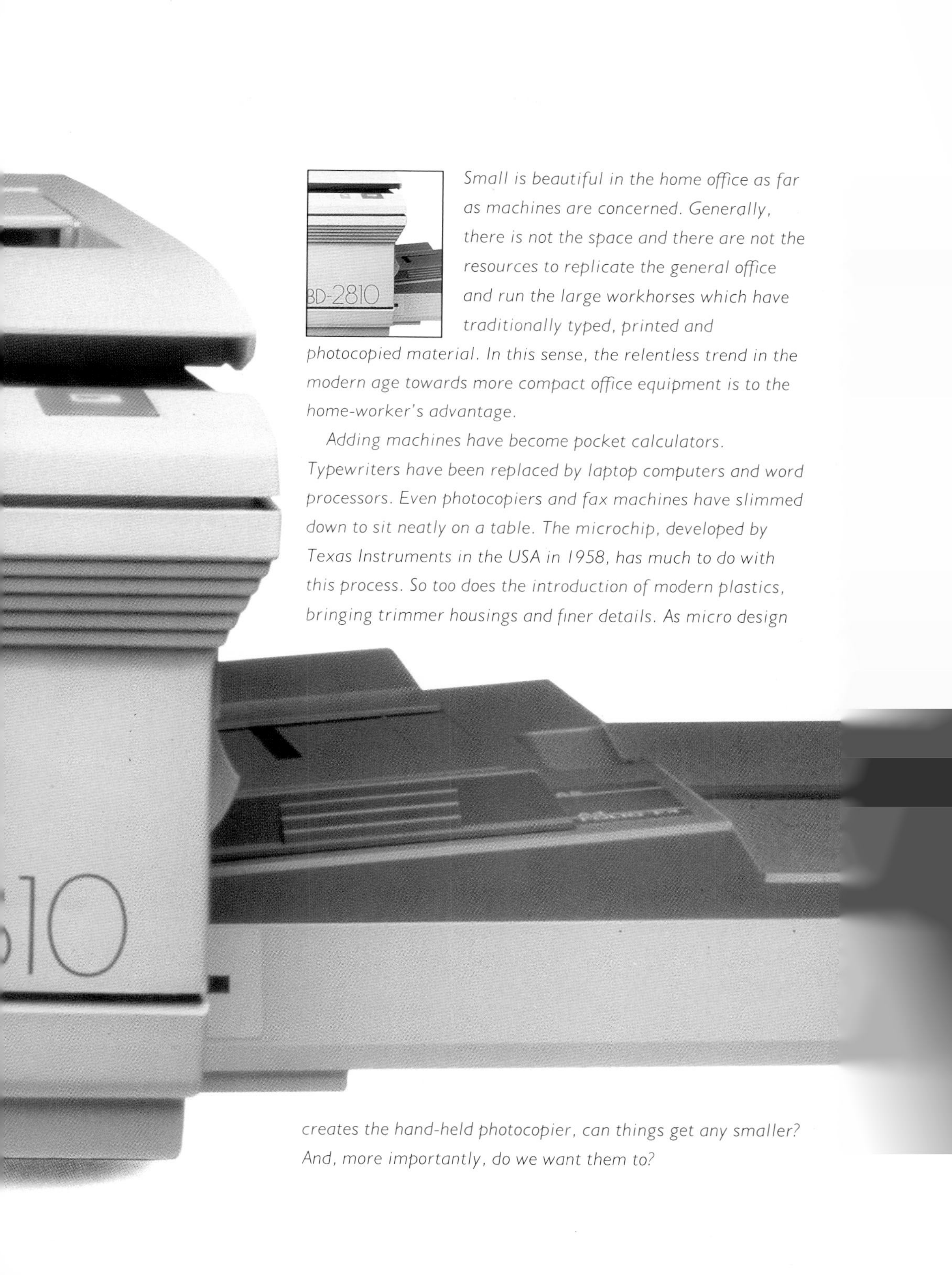

Small is beautiful in the home office as far as machines are concerned. Generally, there is not the space and there are not the resources to replicate the general office and run the large workhorses which have traditionally typed, printed and photocopied material. In this sense, the relentless trend in the modern age towards more compact office equipment is to the home-worker's advantage.

Adding machines have become pocket calculators. Typewriters have been replaced by laptop computers and word processors. Even photocopiers and fax machines have slimmed down to sit neatly on a table. The microchip, developed by Texas Instruments in the USA in 1958, has much to do with this process. So too does the introduction of modern plastics, bringing trimmer housings and finer details. As micro design creates the hand-held photocopier, can things get any smaller? And, more importantly, do we want them to?

SHOLES AND GLIDDEN TYPEWRITER
Christopher Lathan Sholes and C Glidden for E Remington & Sons 1874
The first practical typewriter of the modern age could print only capital letters and the typist was unable to see what was being typed. But many important features were established – most notably the type-bar principle whereby type is moulded on to the end of a pivoted key which strikes the paper. After a slow start the machine grew more popular. In 1881 the New York YWCA bought six typewriters and began typing lessons with a class of eight women. Five years later there were an estimated 60,000 typists employed in the USA. The influential Sholes and Glidden machine has stencilled floral decoration to indicate its intended use by women.

MODEL 66 OFFICE DUPLICATOR

Raymond Loewy for Gestetner 1929

Loewy was given five days and $3,000 by English manufacturer Sigmund Gestetner to redesign the company's 1923 duplicator, a machine with exposed moving parts which quickly became clogged with ink and dust. He decided to solve the problem by housing the duplicator in a sleek Bakelite moulding. The new design was not produced until 1933, but it then rapidly set a trend for increasing visual simplicity and concealed parts in office equipment.

ADDING MACHINE

Contex 1948

The adding machines developed for commerce in the nineteenth century were large, heavy and cumbersome. This stylish Danish model marks the beginning of the process of simplification and miniaturization which led to the plastic pocket calculators of the 1970s and 1980s. It has a phenolic-laminate base-plate, urea-formaldehyde buttons and a Bakelite housing. Its compact, curved shape expresses the design potential of plastics as well as prefiguring the form which new technology would bring to office products.

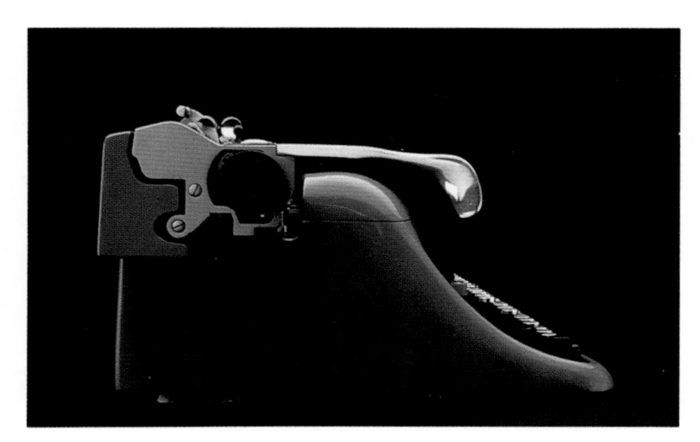

LEXIKON 80 TYPEWRITER
Marcello Nizzoli for Olivetti 1948
This typewriter proved a significant model in Olivetti's post-1945 drive to build a world reputation for design excellence. It was designed by Marcello Nizzoli, who trained as a painter, and engineered by Guiseppe Beccio. Its sculptural shell body is influenced by the aerodynamic form of a bird in flight – a point highlighted in the advertising (*right*). Olivetti was no longer just manufacturing products but was consciously out to promote a specific image. Nizzoli, who worked first in the company's advertising office, was the ideal man.

IBM EXECUTIVE ELECTRIC TYPEWRITER
Eliot Noyes for IBM 1959
A landmark in IBM's programme to give its products identity, this machine, with its formed curves adapted to appeal to women, was very successful. IBM had made electric typewriters since 1933, but by the 1950s it was envious of Olivetti's design reputation so Noyes was enlisted.

IBM 72 SELECTRIC TYPEWRITER

Eliot Noyes for IBM 1961

The IBM 72 was the world's first golfball typewriter – but this major innovation had little impact on design. Exchange of type-bars for a sphere bearing every character meant the carriage no longer moved, yet styling stayed true to the 1959 IBM Executive.

VALENTINE PORTABLE TYPEWRITER

Ettore Sottsass for Olivetti 1968

The first portable typewriter with integral case was aimed initially at the student market but has since become an industrial-design classic. Its sculpted plastics form, confident detailing and bright red colour reflect the optimism of the 1960s. Sottsass, a major influence in Italian design, began working as a consultant to Olivetti in 1957. He was assisted by British designer Perry King.

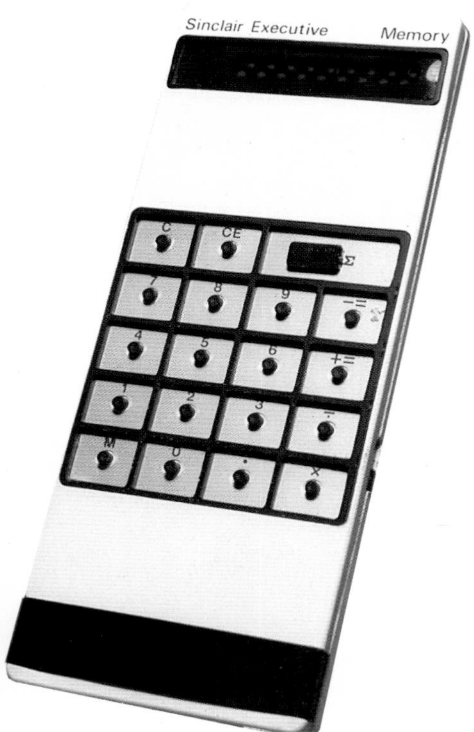

SINCLAIR EXECUTIVE ELECTRONIC CALCULATOR
Clive and Iain Sinclair for Sinclair Radionics 1972

The world's first pocket calculator, this British-designed best-seller weighed just 60g ($2\frac{1}{2}$oz), measured 14 x 6cm ($5\frac{1}{2} \times 2\frac{1}{4}$in), and was only 1cm ($\frac{3}{8}$in) thick. Yet it had more features than many larger machines. It was developed by former technical journalist Clive Sinclair, who founded Sinclair Radionics in 1963, and his industrial-designer brother.

DIVISUMMA 28 CALCULATOR
Mario Bellini for Olivetti 1973

Although Sinclair made the technical breakthrough, Olivetti was poised to exploit the move towards smaller, more compact calculators in design terms. This stylish, ergonomically sound desktop electronic calculator with printer makes the Sinclair machine look crude in comparison. Bellini's ability to give machines a tactile personality is evident in the angle of the injection-moulded ABS plastics body. The keyboard comprises a rubber skin stretched over articulated push buttons.

ET44 ELECTRONIC CALCULATOR

Dieter Rams and Dietrich Lubs for Braun

1977

For many people, this is the classic pocket calculator of the twentieth century. Designed in the actual shape of a pocket with a straight top and curved base, it symbolizes the Braun (and German) quest for technical perfection, engineering precision and stylistic purity. The famous maxim of Dieter Rams – that the main role for the designer is to reduce the visual clutter in the world – is admirably expressed by a machine whose elements have been relentlessly reduced to the essentials, yet it is so easy and enjoyable to use. The Braun ET44 calculator comes with its own slim plastics case.

TES 401 TEXT-EDITING SYSTEM
Mario Bellini for Olivetti 1978
One of the earliest word processors the TES 401 was made accessible by intelligent design. The technical aspects of the machine suggested a bulky, cumbersome form but Bellini separated the mass into three distinct parts. He also recessed the keyboard so it was flush with the surface, and varied the machine's textures and colours to add visual interest. The body and lid are in ABS plastics.

PCW 9512 WORD PROCESSOR
Amstrad 1988
Amstrad – the brainchild of British electronics entrepreneur Alan Sugar launched its first word processor in 1985. It proved a boon to home-workers and undercut most electronic typewriters in price. The versatile PCW 9512 took the concept further by adding more sophisticated features and replacing the dot matrix printer with a daisywheel.

APPLE MACINTOSH
frogdesign for Apple Computer 1984
The Apple Mac personal computer has proved to be one of the most important and original innovations of recent times. Its 'friendliness' in operation and appearance have set an industry standard, while the flexibility of its graphics facility has enabled even the smallest business to compete.

GRIDLITE LAPTOP COMPUTER
Winfried Scheuer for Grid Systems 1986
This laptop computer, with grooved case small enough to slip under your arm, was designed by Winfried Scheuer of ID Two. The angle of the screen can be adjusted to suit different lighting conditions. Its basic configuration is similar to a Grid Systems computer designed in 1980 by ID Two founder Bill Moggridge.

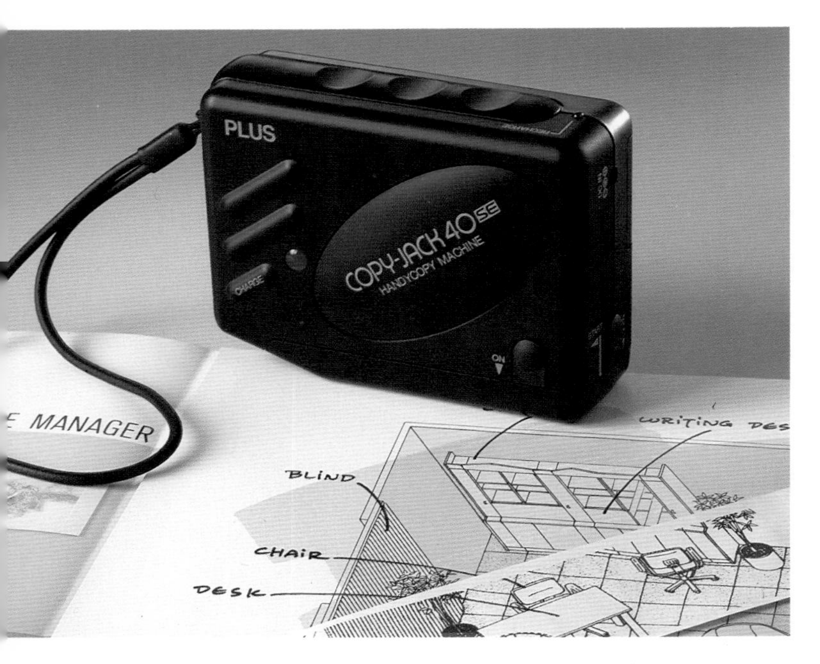

COPY-JACK MINI PHOTOCOPIER

Plus Corporation 1987

The smallest photocopier in the world was developed in Japan. Incredibly, it is hand held, yet its resolution – as demonstrated by the copied architectural layout shown here – is a match for many far larger machines. Narrow strips of thermal copy paper are fed into the Copy-Jack, which works without toners or chemicals and is mains or battery operated.

AGENDA ELECTRONIC ORGANIZER

BIB Design Consultants for Microwriter Systems

1988

The high-technology alternative to the Filofax or Lefax personal organizer, Agenda is a powerful micro-sized word processor. It is made of ABS plastics with a membrane keyboard, and includes many sophisticated features to manipulate information. This high-style accessory designed by BIB won a British Design Council Award in 1990.

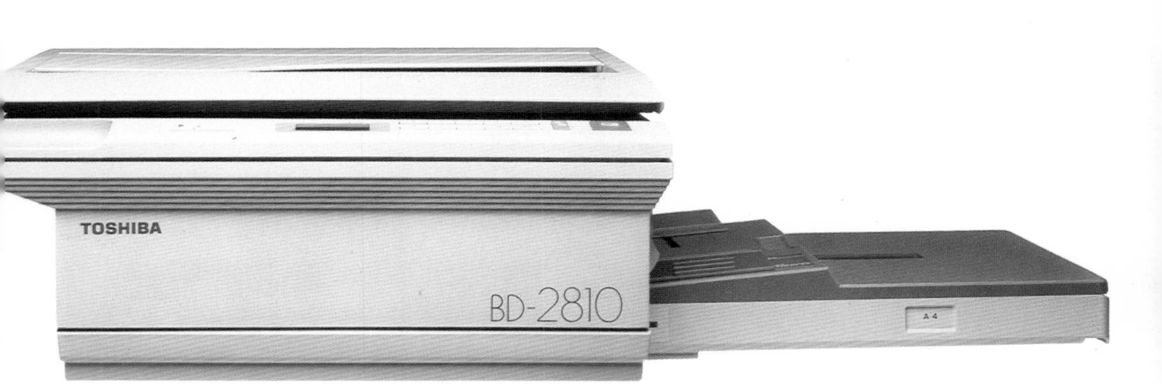

2810 TABLETOP PHOTOCOPIER

Toshiba 1989

Technology plays such a large part in office life that recent design trends have been to simplify processes for the end user and not to revel in complexity for its own sake. As photocopiers have slimmed in size, Toshiba of Japan has emerged as a leader, producing machines for the home-worker which are accessible and effective. The 2810 tabletop copier is a good example, unobtrusively styled with a simple control console.

CANON FAX 80

Canon 1989

The Americans developed the facsimile machine because of unreliable postal services; the Japanese did the same because of the pictorial nature of their language, which made handwritten notes more useful than a telex. It is the Japanese who are now producing the best fax machines for the home office: the sleek Canon Fax 80 doubles as a phone, trebles as a photocopier and is no wider or weightier than a portable typewriter.

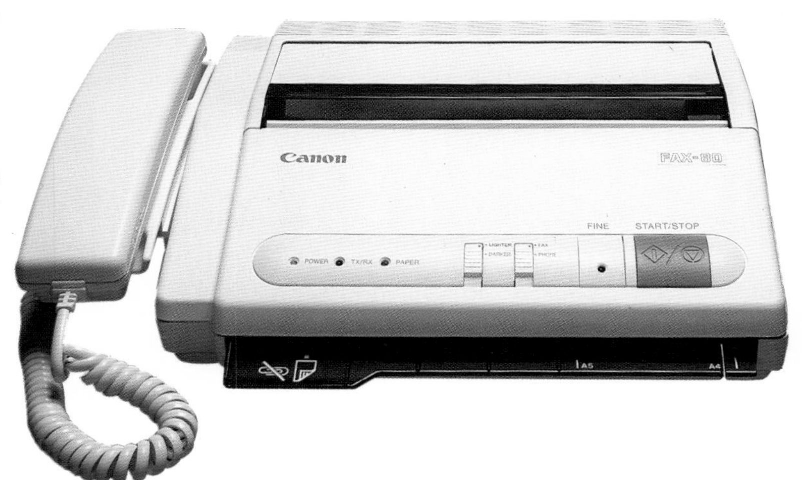

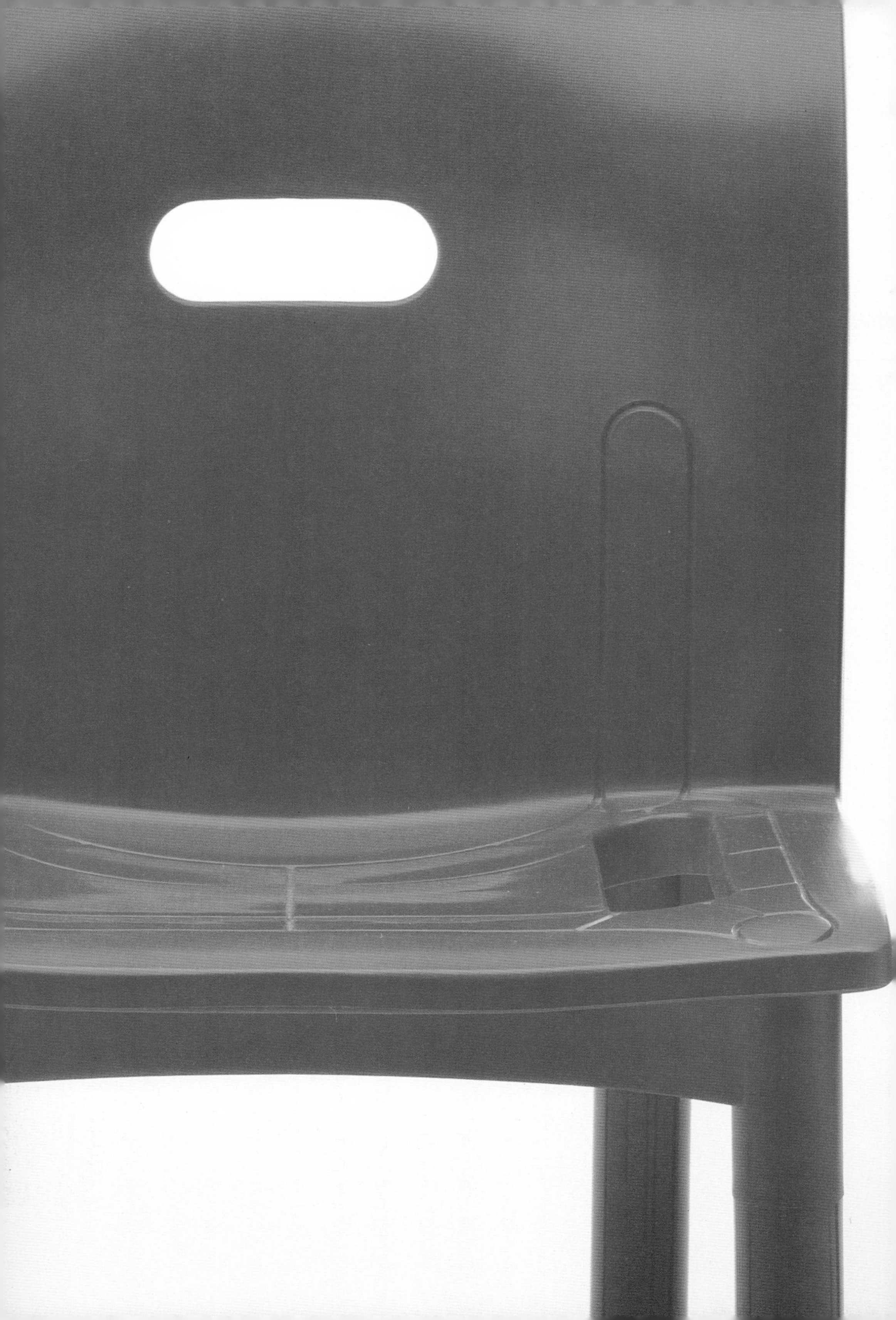

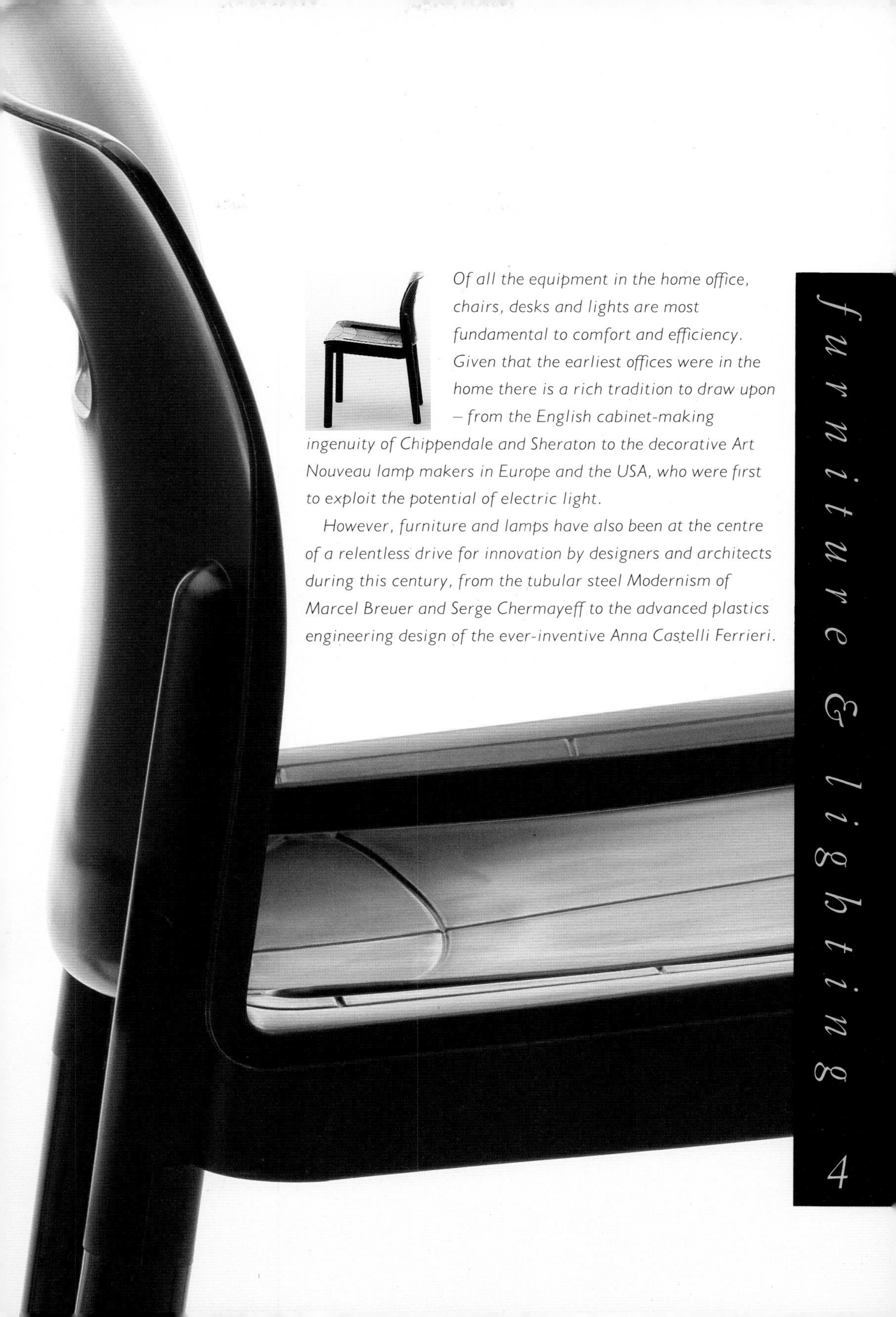

furniture & lighting 4

Of all the equipment in the home office, chairs, desks and lights are most fundamental to comfort and efficiency. Given that the earliest offices were in the home there is a rich tradition to draw upon – from the English cabinet-making ingenuity of Chippendale and Sheraton to the decorative Art Nouveau lamp makers in Europe and the USA, who were first to exploit the potential of electric light.

However, furniture and lamps have also been at the centre of a relentless drive for innovation by designers and architects during this century, from the tubular steel Modernism of Marcel Breuer and Serge Chermayeff to the advanced plastics engineering design of the ever-inventive Anna Castelli Ferrieri.

WRITING DESK

Henry van de Velde 1899

This dynamically shaped desk was one of the pieces which brought Belgian Art Nouveau pioneer Henry van de Velde to European prominence. A major design theorist, he did much to lay the groundwork for Modernism. The desk embodies his belief that 'A line is a force like all elemental forces. It takes force from the energy of the person who has drawn it.'

TABLE LAMPS

Loetz c. 1900

These table lamps – one bell-shaped, the other with a globe shade – are typical of the designs of Austro-Hungarian company Loetz, which specialized in lustrous iridescent Art Nouveau glass. Loetz was among the first European Art Nouveau manufacturers to exploit electric light to create stylish objects for the affluent.

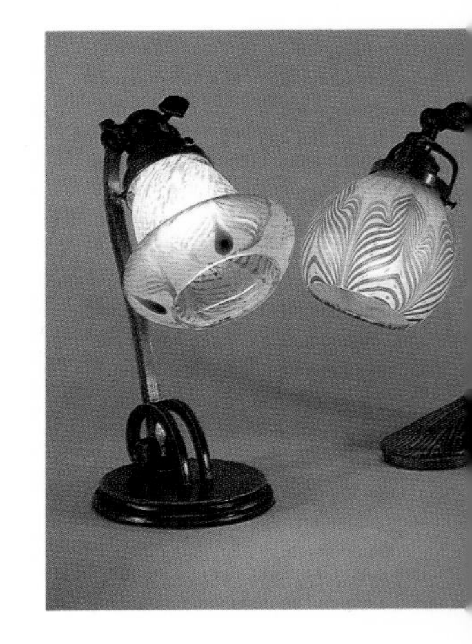

JOHNSON WAX FURNITURE

Frank Lloyd Wright for Steelcase 1936–9

When one of North America's most famous modern architects Frank Lloyd Wright designed a spectacular new administration building for S C Johnson & Son in Racine, Wisconsin, he created custom desks and chairs to match the environment. The classic pieces which resulted echoed the breathtaking curves and columns of the architecture and became a forerunner to modern office-systems furniture. Craftsmen from US office manufacturer Steelcase worked with Wright on the futuristic cast-aluminium and magnesite pieces. The desk has a two-tier working surface and swing bins instead of drawers; the chair (*see detail*) has just three legs and encourages good posture. The furniture enhanced Wright's aim that the Johnson Wax building should be 'as inspiring a place to work as any cathedral ever was in which to worship'.

CESCA CHAIR
Marcel Breuer for the Bauhaus with Mannesmann Steel 1928
Hungarian-born architect and furniture designer Marcel Breuer devoted much of his time as a tutor at the Bauhaus to developing a new aesthetic for tubular steel. His much admired and reproduced Cesca cantilever chair has been particularly popular for home-office use because it combines, with great dexterity, the industrial Modernism of nickel-plated steel tubing with an artisan-inspired cane seat and back. The result is a design of refined economy which expresses the values of the Machine Age, yet works well in a domestic setting.

SP3 ARMCHAIR
Serge Chermayeff for PEL 1931
Inspired by Marcel Breuer, Mart Stamm and German-made tubular-steel furniture in London's Strand Palace Hotel, two directors of British company Tube Investments formed a subsidiary called PEL (Practical Equipment Ltd) which did much to advance modern design in Britain in the 1930s. This cantilever spring armchair was designed by Russian-born architect and designer Serge Chermayeff. The frame is formed by a continuous steel tube, creating a balanced and harmonious design which was specified by the BBC.

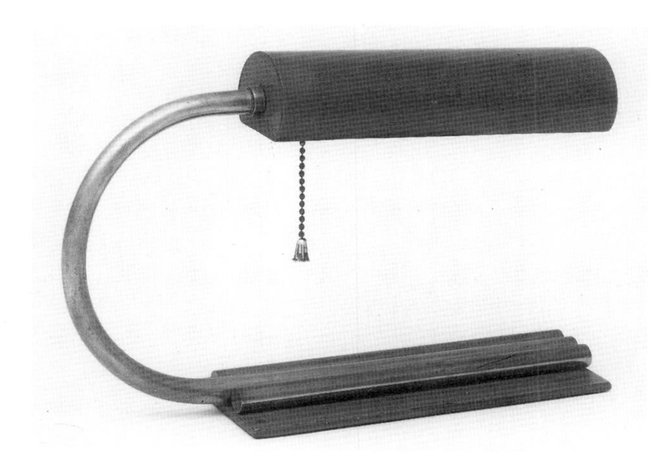

DESK LAMP
Gilbert Rohde for Mutual Sunset Lamp Manufacturing Company
c. 1933
This simple, streamlined desk lamp belongs to the same period as Donald Deskey's Art Deco interiors for the Radio City Music Hall in New York, and reflects European influence on the bold forms which emerged in lighting in the USA during the 1930s.

TABLE LAMP
Mario Fortuny c. 1929
Venice-based photographer and fashion designer Mario Fortuny emerged as one of the most creative talents from the maelstrom of new design ideas in the early years of this century. His interest in stage design and lighting led him to create a series of table lamps with real presence. The original of this model is now in the Fortuny Museum in Venice, while a reproduction has been issued in Britain. The black-lacquered, chromed-steel arc frame stands on an oak base. The reflector is in chromed aluminium with a white enamel interior to enhance light reflection.

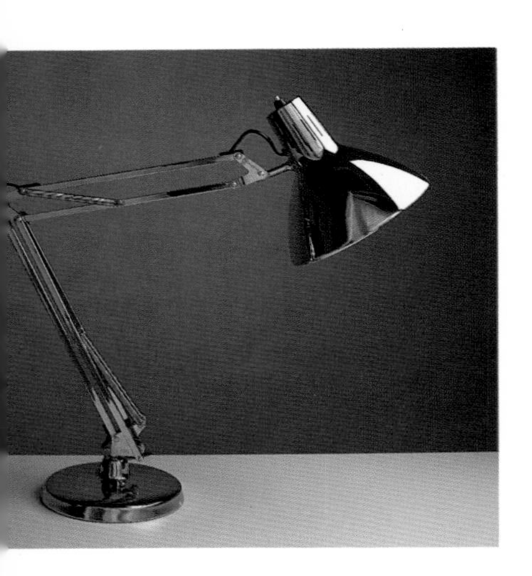

LUXO L-1 TASK LAMP

Jac Jacobsen for Luxo

1937

Derived from the classic British Anglepoise lamp of 1934, the Luxo is an adjustable-arm cantilever task light which has enjoyed huge success the world over. Norwegian manufacturer Jac Jacobsen took out a patent on the Anglepoise after seeing the lamp on a trip to England. With modification, this became the Luxo, which enjoyed success throughout Scandinavia before its American introduction in 1951. The Luxo can be attached to wall, desk edge or base; one leading designer even mounted his model in a block of granite. A functional workhorse, it spans desks and floors, giving light exactly where it is needed.

606 UNIVERSAL SHELVING SYSTEM

Dieter Rams for Vitsoe

1960

When German master of Functionalism Dieter Rams turned his attention to home-office shelving, the result was a product of scrupulous and unsurpassed versatility and attention to detail. The shelves are in three finishes – epoxy powder-coated steel, lacquered wood and veneered wood – with a support structure in extruded anodized aluminium. The system can build from a couple of shelves into a complete storage system with add-on cabinets.

DKR CHAIR
Charles Eames for Herman Miller 1951
One of a series of classic chairs created by Eames for Herman Miller in the 1950s, the DKR shows this influential designer's special skill in making connections between different materials and forms. The base is chromium-plated steel rod, while the upholstery is cut away to reveal a skeletal lattice wire shell. The feet are self-levelling: there is a slight springiness as you settle into the chair. Its tensioned struts are influenced by aircraft technology rather than furniture design – Eames worked on aircraft during World War Two. He seemed to capture a new mood in furniture, and his chairs have been described as belonging to the occupant, not the building, unlike the work of other modern architects.

ROUND TABLE
Florence Knoll for Knoll International 1961
Despite a formidable training and a pioneering role in introducing modern furniture classics to corporate America, Florence Knoll modestly dismissed her designs as 'fill-ins'. But many have stood the test of time: this table/desk in the KnollStudio Collection has a chrome-finished steel base and a round white Italian-marble top. It is also available in walnut, teak and rosewood veneer.

KEVI CHAIR

Jorgen Rasmussen for Kevi 1970

Natural in style and ergonomic in design, the classic Danish Kevi office chair is made of coloured moulded plywood.

PLIA CHAIR

Giancarlo Piretti for Castelli 1969

Piretti's economical folding chair, made using a toughened form of cellulose acetate, has inspired many imitations. It goes with the Platone desk.

GF 40/4 CHAIR

David Rowland for General Fireproofing 1964

In vinyl-covered sheet metal with a tubular-steel frame, this brilliant US design stacks 40 high in just 1.25m (4ft).

MODULAR DRAWER SYSTEM

Simon Fussel for Kartell 1974

Produced by Kartell in glossy ABS plastics, this Italian system is ideal for the home-worker. One drawer is attached to another by means of four small cylindrical plugs building up into a versatile storage system. In 1988 Olaf von Bohr designed trestles and tops.

4807 STACKING CHAIR
Anna Castelli Ferrieri for Kartell 1985

Launched at the 1985 Milan Furniture Fair here is a design which expertly met the challenge of mass producing an injection-moulded plastics chair capable of stacking easily. Kartell art director Anna Castelli Ferrieri inserted two slots in each seat, through which the rear legs of another chair could fit to enable smooth stacking (*see detail*); usually space is wasted by setting the rear legs outside the chair in the stacking arrangement. The chair is brilliant in other ways: the slot in the backrest gives a firm grip in handling, while sink marks in the polypropylene, caused by extra ribbing underneath to strengthen the chair, have been turned into surface decoration that is integral to the form of the piece. Kartell won a Compasso d'Oro award for the design in 1987. Highly comfortable to use, it has subsequently been produced with armrests.

TALO DESK
Massimo Scolari for Giorgetti 1989
The classical tradition of furniture-making exerts fresh influence on contemporary design for the home office. The visual and technical ingenuity of eighteenth-century English cabinet-makers Chippendale and Sheraton has informed recent work, such as this desk by Italian designer Massimo Scolari. Developed around a central column of hand-worked solid maple with a work-surface of cherry or walnut, a special mechanism allows independent rotation of the work-surface and drawer. Emerging from the column is a brass and copper lamp which has an old-fashioned light-pull. In a sense we have come full circle – but the traditional aesthetic is achieved by combining craftwork with modern technology.

SECRETAIRE
Richard Sapper for Unifor 1989

As the boundaries between home and work become progressively blurred, designers are returning to the aesthetic forms of the earliest offices in the home – the drawing rooms and libraries which existed before commerce left the domestic setting. The Secretaire is unashamedly inspired by antique furniture yet it is designed with modern electronics very much in mind. Inside the refined timber cabinet, which rests on castors, there is space (and a power point) for a lamp, your personal computer, plus room for ancillary storage. You can work at the Secretaire standing or sitting: a counter-balance mechanism allows the top cabinet to rise and fall. Everything folds away after use so the unit can revert to being a domestic piece of furniture. Clearly, this is a marked departure from the conventional office desks which have dominated this century. It looks forward to an era of increased home-working – and back to an age of inspiring classical furniture.

MARIO BELLINI

Leading Italian industrial-design consultant Mario Bellini (b. 1935) was among the first designers to realize that the microchip had rendered the old adage of 'form follows function' irrelevant. His work has consequently explored the issues of product identity, often with strongly sculptural forms.

Bellini graduated in architecture from Milan Polytechnic in 1959. He went on to create much pioneering design work for Olivetti, including the **Divisumma 28 electronic calculator** of 1973 and the **TES 401 Text-Editing System** five years later. Lighting for Erco, Flos and Artemide, and furniture for Cassina and Vitra, also bear his intelligent and tactile hallmark.

Bellini became editor of the influential *Domus* magazine but he is less of a theorist and polemicist than other leading figures in the Milanese design hothouse. His achievements were honoured with a major exhibition at New York's Museum of Modern Art in 1987.

LASZLO BIRO

Hungarian inventor Laszlo Biro (1889–1985) gave his name to the Biro pen. Although details of his early life are sketchy it is known that he served as an officer in the Hungarian Army during World War One and engaged his fellow officers in endless debate about pens.

Biro was also a tireless inventor. By the time he took out his first patent on the ballpoint pen in 1938, he had already patented some 30 minor inventions.

During the 1930s Biro edited a cultural magazine in Hungary. While visiting the print shop he noticed the advantages of instant-drying ink. With his chemist brother George, Biro developed the first commercially successful ballpoint; it relied on a specially formulated ink being delivered to the ball by capillary action.

As a fervent Communist, Laszlo Biro was forced to flee Hungary as the Nazis stormed Europe. He went first to Paris and then to Argentina. He filed an Argentine patent on the ballpoint pen in 1943 but as a penniless refugee was forced to sell the rights. An English entrepreneur named Martin, working in conjunction with the Miles Aircraft Company, began manufacturing the ballpoint in Britain while American pen companies Faber and Eversharp acquired the North American rights.

The Biro pen greatly assisted airmen preparing charts at high altitude during World War Two. When it went on general sale after 1945, it proved an instant success, ensuring Biro a place in modern consumer history.

MARCEL BREUER

One of the most celebrated students of the Bauhaus, Hungarian furniture designer and architect Marcel Breuer (1902–81) is best known for successfully exploring a new aesthetic for tubular-steel furniture. Breuer studied at the Vienna Academy of Art and at the Bauhaus.

Between 1924 and 1928 he ran the furniture workshops at the Bauhaus and developed his famous cantilever tubular-steel chairs. He then set up his own design studio in Berlin before fleeing the Nazis. He went first to England in 1934, where he collaborated with the architect F R S Yorke and designed furniture for Jack Pritchard's Isokon company, and then to the USA in 1937, where he teamed up with Gropius and later founded his own practice.

At Gropius's invitation, Breuer joined the Department of Architecture at Harvard University and buildings became his prime focus. However, it is for his influence on furniture design that he is best remembered.

ANDY DAVEY

British industrial designer Andy Davey (b. 1962) trained at Worthing College of Art and the Royal College of Art.

Upon graduation in 1987 he joined London product-design consultancy Wharmby Associates, where he designed the **Taurus telephone** for Browns Holdings, which won a prestigious Silver Award from the Design and Art Directors Association in 1989.

In 1990 Davey founded his own design consultancy, TKO, in Tokyo.

HENRY DREYFUSS

American industrial designer Henry Dreyfuss (1904–72) belonged to the select band who laid the foundations of the international design-consultancy business during the 1930s. Together with Raymond Loewy, Walter Dorwin Teague and Norman Bel Geddes Dreyfuss defined the role of industrial design in American industry.

He was born into a family which specialized in prop and costume hire and went into the theatre at the age of 16. He opened his first industrial-design office in 1929 and he became one of the USA's most powerful and influential designers, creating vacuum cleaners for Hoover, televisions for RCA, tractors for John Deere and plane interiors for Lockheed.

Dreyfuss, Loewy and their peers have been termed stylists, creators of the luxurious images which enabled American industry to woo new customers in the post-Depression 1930s. But Dreyfuss was at heart a Functionalist. He avoided excessive elaboration, and his fascination with anthropometry resulted in a book *Designing for People* which laid the groundwork for industrial design based on scientific measurement and the development of ergonomics. His classic **1937 Bell 300 telephone** demonstrated his talent for restrained design that was appropriate to human needs.

CHARLES EAMES

Best known for a series of chairs designed for Herman Miller in the 1950s, American furniture designer Charles Eames (1907–78) has exerted a major influence over US design education and practice.

Eames was born in Saint Louis, Missouri, and worked as a technical draughtsman before becoming an architect.

He opened his own design office in 1930 in the mid West. A decade later he gained national attention when he and fellow architect Eero Saarinen won a home-furnishings competition organized by the New York Museum of Modern Art. In 1946, Eames became the first designer to have a one-man show at the Museum.

Eames's early furniture explored a combination of moulded plywood and steel rods. Using delicate and mobile forms, his work was characterized by great attention to detail. In 1949, Charles Eames and his wife Ray Eames built a house in Santa Monica entirely from pre-fabricated glass and steel elements. It fascinated American design students and later Eames made a short film about its construction, entitled *Powers of Ten*.

By then his reputation was secure. His 1956 lounge chair and ottoman, designed for film director Billy Wilder, had become a classic; his Herman Miller designs were widely acclaimed; and his incisive forays into lecturing had secured critical acclaim despite a relatively slim body of work.

HARTMUT ESSLINGER

West German designer Hartmut Esslinger (b. 1945) is the founder of Frogdesign, one of the world's leading industrial-design consultancies. Esslinger studied electrical engineering at the University of Stuttgart and industrial design at the College of Design in Gmund.

He established Frogdesign in 1969. His first client was the electronics company Wega Radio. When Wega was bought out by Sony, he gained an important foothold in the Japanese market.

Esslinger opened Frogdesign offices in California in 1982 and in Tokyo in 1986. In the USA, the group is best known for its work for Apple Computers, including the **Apple Mackintosh**. In Europe it has designed office furniture for Konig & Neurath.

Esslinger belongs to the German Functionalist tradition, but his work is tempered with more colourful, quirky elements from the Post-Modernist repertoire.

ANNA CASTELLI FERRIERI

Italian designer and architect Anna Castelli Ferrieri (b. 1918) studied architecture at the University of Milan, where she later worked as a lecturer.

Ferrieri's career has extended from town planning to furniture. Her name is most closely associated, however, with the leading Italian plastics company Kartell. Ferrieri became a consultant to Kartell in 1966 and since 1976 she has been the company's art director. Her achievements there were honoured in 1979 with the coveted Premio Compasso d'Oro for industrial design.

Ferrieri epitomizes the modern generation of imaginative Italian designers. Her 1985 polypropylene **stacking armchair** is indicative of her enduring creative and technical ingenuity.

FLORENCE KNOLL

American furniture and interior designer Florence Knoll (neé Schust) was born in Michigan in 1917. She trained as an architect under Eero Saarinen, Ludwig Mies van der Rohe and at the Architectural Association in London.

She married modern ırniture pioneer Hans Knoll . 1943. Three years later they founded Knoll Associates, later known as Knoll International, a company dedicated to introducing modern furniture design to corporate America.

Hans Knoll's entrepreneurial skills and Florence's design credentials ensured a huge success. Designers including Isamu Noguchi in the USA, Franco Albini in Italy and Pierre Jeanneret in France were commissioned on royalty agreements; Marcel Breuer and Ludwig Mies van der Rohe became famous names via Knoll's post-war domination of the look of American office interiors.

After Hans's death in 1955, Florence Knoll maintained the company's high design standards. She ran the Knoll Planning Unit, drawing on her architectural training, and designed much memorable furniture, although she dismissed her pieces – sofas, cabinets, desks and conference tables – as 'fill-ins'. She remarried but remained design director of Knoll until her retirement in 1965.

Her achievement in introducing Modernist furniture design to American business has been widely acclaimed. She received the Gold Medal for Industrial Design from the American Institute of Architects in 1961.

RAYMOND LOEWY

French-American industrial designer Raymond Loewy (1893–1986) was a pioneering showman, the prime mover in the creation of the international design-

consultancy business.

Loewy was born in Paris and became a captain in the French Army. In 1919 he went to New York and gained work as a fashion illustrator and display designer. In 1927 he founded his own design studio.

In 1929 Loewy was given five days to redesign a **Gestetner Office Duplicator Machine**. The result – a clever styling exercise – has become one of the milestones of design in industry and helped to establish the concept of the industrial designer in the USA.

Together with Henry Dreyfuss, Walter Dorwin Teague and Norman Bel Geddes, Loewy is associated with the rise of styling in the 1930s: a conscious streamlining of form sneered at by the purists but which nevertheless ensured design's place in the production process.

Loewy proved to be the most successful of his peer group: by 1946 he had 75 international clients in a wide range of industries. Much of his work became icons of contemporary consumer life – from Lucky Strike cigarette packaging to the Studdebaker car. He was the first designer to make the cover of *Time* and his influence extended into the Kennedy era: he designed the interiors for President Kennedy's personal Boeing 707.

Loewy had detractors, who dismissed him as a flamboyant stylist. But his work revealed a profound interpretation of the commercial worth of attractive aesthetics and an inspirational awareness of functional values.

MICHELE DE LUCCHI

One of the best known members of the Italian Memphis group, Michele de Lucchi (b. 1951) graduated in architecture from Florence University in 1975.

While a student he had founded the avant-garde

Gruppo Cavart, which worked on a number of radical projects in architecture, design and film. As a graduate, he became an assistant professor in his old department and also taught at the International Art University of Florence.

In 1978, however, de Lucchi moved to Milan to become a design consultant in close association with Ettore Sottsass. They worked in the avant-garde group Studio Alchimia and collaborated on the launch of Memphis in 1981, for which de Lucchi subsequently designed many pieces.

Under Sottsass's supervision, de Lucchi designed the Icarus office-furniture system for Olivetti and the interiors for 50 Fiorucci stores worldwide during the 1980s. He now works for a number of Italian manufacturers, including Artemide, Kartell and Fontana Arte.

ENZO MARI

Italian designer Enzo Mari (b. 1932) studied at the Brera in Milan in the early 1950s. In 1963 he coordinated the Italian group Nuove Tendenz (New Tendencies) and in 196 was responsible for an exhibition of optical, kinetic and programmed art at the Zagreb Biennale.

In 1972 he took a major role in the exhibition *Italy: The New Domestic Landscape* at the Museum of Modern Art in New York as his experimental work in exhibitions, graphics, publishing and industrial products thrust him to the forefront of Italian design.

Mari's international reputation has been built on his ability to bring an artist's

perception to the design of utilitarian objects. His concern has been to give products new forms and meanings, and he has taught his design methods at Milan Polytechnic.

Best known of Mari's ndustrial work is his range of roducts for Danese in the 60s, including a number of sk accessories. He has also signed furniture for Driade d Gabbinanelli. He has been warded the Compasso d'Oro, Italy's premier industrial-design prize, on three occasions and he served as president of the Italian Association of Industrial Design from 1976 to 1979.

MARCELLO NIZZOLI

Italian industrial designer Marcello Nizzoli (1887–1969) was born in Boretto and studied at the School of Fine Arts in nearby Palma. He first came to prominence as a painter, but it was for the sculptural lines of two of Olivetti's most important pioneering typewriters that he was widely acclaimed.

Nizzoli, who worked variously as a poster artist and an exhibition and fabric designer, joined Olivetti in 1938 to work in the company's advertising office. He went on to design Olivetti's **Lexikon 80 typewriter** in 1948 and the Lettera 22 portable typewriter in 1950.

Nizzoli hid the mechanical components within sophisticated curved shells, a technique he repeated with his Mirella sewing machine for Necchi in 1956.

ELIOT NOYES

One of the most important and influential industrial designers in the USA during this century, Eliot Noyes (1910–77) was responsible for bringing the Bauhaus to big business.

Noyes studied architecture at Harvard University in the 1930s,where he met Bauhaus refugees Walter Gropius and Marcel Breuer. The writing of Le Corbusier further stimulated his enthusiasm for European Modernism and he reacted against his Beaux-Arts training.

An introduction by Gropius enabled Noyes to become Director of the Department of Industrial Design in the New York Museum of Modern Art, a post he held just before and after World War Two. In

the late 1940s he worked for Norman Bel Geddes before setting up on his own as an industrial designer and architect.

Noyes immediately made his name as a consultant to IBM. Thomas Watson, IBM's founder, hired him to give the company a corporate identity. Noyes brought in such Modernist luminaries as Marcel Breuer to work on buildings, Charles Eames to work on products, and Paul Rand to design graphics, mainly devoting his own time to orchestrating the rebirth of IBM. In 1956 he was formally appointed corporate design director. American big business was so impressed that he was appointed first by

Westinghouse (in 1960) and then by Mobil Oil (in 1964) to perform similar feats.

Noyes introduced an uncompromising design philosophy to the highest levels of American commerce and industry. His legacy was a significant raising of standards with a greater emphasis on design integrity.

GIANCARLO PIRETTI

Italian designer Giancarlo Piretti (b. 1940) trained as an art teacher at the Istituto Statale d'Arte in Bologna. Upon graduation he taught industrial design there for seven years before spending 12 years as a designer and then director of research and design for the Italian furniture manufacturer Castelli. Among his notable designs are the **Plia chair (1969)**, Platone desk (1971) and Vertebra seating system (1979).

Vertebra was designed in collaboration with Emilio Ambasz. The two teamed up again to work on the Logotec spotlight system for Erco in 1980.

In 1988 the Piretti Collection, a range of 50 chairs, was launched in the USA by Krueger.

DIETER RAMS

The most important industrial designer of post-war Germany, Dieter Rams (b. 1932) was apprenticed as a joiner and later studied architecture and design at Wiesbaden School of Art in the town where he was born.

He worked initially for the architectural firm of Otto Apel and collaborated with the American designers Skidmore Owings and Merrill, who were involved in US consulate buildings in West Germany at the time.

In 1955, Rams was appointed chief designer at Braun. He developed an austere style which combined clarity and simplicity with sculptural presence. By 1959 many Braun products were on display in the New York Museum of Modern Art. Nobody better exemplifies the effect of the Modern Movement on manufacturing industry.

His approach is felt in the home-office sphere, from Braun clocks and fans to a shelving system for Swiss furniture company Vitsoe. The **Braun ET44 electronic calculator**, which Rams designed with Dietrich Lubs in 1977, is a classic.

Since 1981 Rams has been professor of industrial design at the Hochschule für Bildende Künste in Hamburg. He combines educational commitments with a gruelling lecturing and consultancy schedule. Braun is today owned by the American company Gillette so Rams is extending his principles into US industry.

ETTORE SOTTSASS

One of the most unpredictable and outstanding talents of post-war Italian design, Ettore Sottsass was born in 1917 in Innsbruck, Austria.

He studied architecture at Turin Polytechnic and opened a design office in the city in 1946. Early work on housing, interiors and small decorative products was followed by a pioneering consultancy to the electronics division of Olivetti.

The collaboration, which began in 1957, has produced such classics as the famous red **Valentine portable typewriter**, the Tekne and Praxis electric typewriters and the Synthesis office-furniture range.

As a design guru, Sottsass has exercised a major influence over the Italian industrial scene. In the 1960s he became interested in Pop Art. In the 1970s he carried out commissions for firms such as Poltranova and Alessi

as well as contributing furniture designs to avant-garde group Studio Alchimia's ironically named Bauhaus Collection.

But in 1981 he broke away to form a radical design movement – Memphis – which became a focus for the visual and intellectual irreverence which have always been features of his work. Memphis overturned Modernist good taste which had dominated Milanese design, and Sottsass became the figurehead of an international Post-Modernist movement in interiors and products.

Today, with the influence of Memphis on the wane, Sottsass works with architects Aldo Cibic, Matteo Thun and Marco Zanini in Sottsass Associati, the design consultancy he formed in 1980.

FRANK LLOYD WRIGHT

The most legendary name of twentieth-century American architecture, Frank Lloyd Wright (1867–1959) extended his extraordinary artistic and technical grasp to the details of interiors: furniture, light fittings and tableware.

Wright learnt his trade as a Wisconsin engineering

student and then in the Chicago architectural office of Adler and Sullivan. In 1893 he set up his own office and by 1900 his career had taken off.

His influence has been so powerful that a diverse range of designers, from advocates of the Modern movement in the 1940s to Post Modernists in the 1980s, have claimed that his ideas predated theirs.

But Frank Lloyd Wright was clearly in a league of his own, as demonstrated by the famous Johnson Wax building of 1939, for which he also designed the desks and chairs. Wright was an extraordinary synthesizer of the best building practice in the USA, Japan and Europe and created some of the most remarkable buildings and furniture designs of the twentieth century.

Page numbers in *italic* refer to the illustrations

accessories, 18–19, 35–47
AEG, 19, *32*, *36*
Amstrad, *56*
Apple Computers, 20, *57*
Arad, Ron, 12, *12*

Bandai, *46*
Bauhaus, 12, 20, 64
Behrens, Peter, 19, *36*
Bell, Alexander Graham, 8, 14
Bellini, Mario, 20, *54*, *56*, 72
BIB Design Consultants, *58*
Bich, Marcel, 17
Bigaglia, Pietro, 18
Biro, Laszlo, 17, *29*, 72
Boyton, Charles E, *39*
Braun, 35, *40*, *55*
Breuer, Marcel, 11, 61, *64*, 72–3

Calendox, *38*
Canon, *59*
Carwardine, George, 13, *13*
chairs, 11, 61, *63*, *64*, *67–9*
Chermayeff, Serge, 11, 61, *64*
Colombo, Joe, 11, *11*
Conran Associates, 19, *19*
Contex, *51*
Cousins Associates, *15*

Davey, Andy, *33*, *73*
Dell, Christian, 12
Deskey, Donald, 12
desks, 9–10, 61, *62*, *70–1*
Diehl, *38*
Dreyfuss, Henry, 15, *28*, 73

Eames, Charles, *67*, *73–4*
electronic machines, 20–1, *49–59*
Ericsson, L M, 14–15, *28*, *30*
Esslinger, Harmut, 74

Faber, Kasper, 18
Fabian, Wolfgang, *17*
Ferrieri, Anna Castelli, 11, *43*, 61, *69*, 74
Fortuny, Mario, 14, *65*
Fussel, Simon, 11, *68*

Gestetner, *9*, *51*

Heiberg, Jean, 15, *28*
Heller, Orlo, *39*

IBM, 20, *52–3*
Igarashi, Takenobu, *47*
Imaizumi brothers, *44*

Jacobsen, Jac, 13, *66*
Jenney, William Le Baron, 9

Kaufmann, Yaacov, *44*
Knoll, Florence, *67*, 75
Krohn, Lisa, 15, *15*

lighting, 12–14, 61, *62*, *65*, *66*
Loetz, *62*
Loewy, Raymond, *51*, 75–6
Lubs, Dietrich, *55*
Lucchi, Michele de, *46*, 76

Mari, Enzo, *41*, 76–7
Maurer-Becker, Dorothee, *42*

Neustadter, Arnold, *39*
Nizzoli, Marcello, *52*, 77
Norman & Hill, *37*
Noyes, Eliot, 20, *52–3*, 77–8

Olivetti, 20, *52–4*, *56*

Parker, J C, *37*
Parker pens, 16, 23, *29*
pens and pencils, 16–18, *26*, *29*, *30*, *32*
Pentel, 17, *30*
Piretti, Giancarlo, 12, *68*, 78
Plus Corporation, *19*, 20, *32*, *44*, *58*

Rams, Dieter, 35, *55*, *66*, 78
Rank Xerox, 20, *20*
Rasmussen, Jorgen, *68*
Remington, E & Sons, 8, *50*
Rohde, Gilbert, *65*
Rowland, David, *68*
Rowlands, Martyn, *31*
Rowney, T & R, 18

Sapper, Richard, 12, 13, *13*, 61, *71*
Scheuer, Winfried, *57*
Scolari, Massimo, 12, *70*
Sholes and Glidden, 8, *50*
Sinclair, Clive, *54*
Solozabal, Olave, *45*
Sottsass, Ettore, 20, *53*, 78–9

Takaichi, Tadao, *46*
Taylor, Frederick, 9
Teague, Walter Dorwin, 12
telephones, 14–15, *24–5*, *27*, *28*, *30–3*
Toshiba, 20, *59*

Underwood, *8*

Vadler, Johann, 19
Van de Velde, Henry, *62*
Veimeister, Tucker, 15, *15*

Wagenfeld, Wilhelm, 12, *27*
Waterman, Lewis Edson, 16–17
Weiss, Reinhold, *40*
Wright, Frank Lloyd, *63*, 79